Vectors in 2 or 3 Dimensions

Other titles in this series

Linear Algebra
R B J T Allenby

Mathematical Modelling
J Berry and K Houston

Discrete Mathematics
A Chetwynd and P Diggle

Particle Mechanics
C Collinson and T Roper

Numbers, Sequences and Series
K E Hirst

Groups
C R Jordan and D A Jordan

Probability
J McColl

In preparation

Ordinary Differential Equations
W Cox

Analysis
E Kopp

Statistics
A Mayer and A M Sykes

Calculus and ODEs
D Pearson

Modular Mathematics Series

Vectors in 2 or 3 Dimensions

A E Hirst

Faculty of Mathematical Studies,
University of Southampton

A member of the Hodder Headline Group
LONDON • SYDNEY • AUCKLAND

First published in Great Britain 1995 by
Arnold, a member of the Hodder Headline Group,
338 Euston Road, London NW1 3BH

British Library Cataloguing in Publication Data
A catalogue record for this book is available from the British Library

ISBN 0 340 61469 2

1 2 3 4 5   95 96 97 98 99

Typeset in 10/12 Times by
Paston Press Ltd, Loddon, Norfolk
Printed and bound in Great Britain by
J W Arrowsmith Ltd, Bristol

Contents

Series Preface

This series is designed particularly, but not exclusively, for students reading degree programmes based on semester-long modules. Each text will cover the essential core of an area of mathematics and lay the foundation for further study in that area. Some texts may include more material than can be comfortably covered in a single module, the intention there being that the topics to be studied can be selected to meet the needs of the student. Historical contexts, real life situations, and linkages with other areas of mathematics and more advanced topics are included. Traditional worked examples and exercises are augmented by more open-ended exercises and tutorial problems suitable for group work or self-study. Where appropriate, the use of computer packages is encouraged. The first level texts assume only the A-level core curriculum.

Professor Chris D. Collinson
Dr Johnston Anderson
Mr Peter Holmes

Preface

Now that vectors are no longer in the core of all A-level mathematics syllabuses, the first time some students meet vectors will be in the first year at university. It seemed a good idea, therefore, to have a book on vectors starting at a very basic level, but also incorporating many of the ideas found in university mathematics courses. This book is written from the geometrical standpoint (indeed, the title was originally to have been *Geometric vectors*), and although applications to mechanics will be pointed out from time to time, and techniques from linear algebra will be employed, it is the *geometric* view which will be emphasised throughout.

Worked examples and exercises are included throughout the text, as it is often through applying techniques in solving problems that a full understanding of the theory behind them is arrived at. At the end of each chapter is a set of exercises designed to put into practice the techniques expounded therein. In some of these end-of-chapter exercises there are *challenge questions* which are aimed at the brave, and, although they may be more difficult to solve than the run-of-the-mill questions, they actually need no more theory than has been explained beforehand, and guidance is given in some cases. Solutions to most exercises are given at the end of the book, although the diagrammatic solutions have been omitted, and solutions to challenge questions are not given. At the end of Chapter 2, a project topic is suggested which would be well within the scope of a good first-year undergraduate (or even someone studying further mathematics at A-level).

The first four chapters are intended to be an introduction to the properties of vectors and the techniques for using them to find vector equations of planes, lines and intersections of planes and planes, planes and lines, and lines and lines. These vector methods are also the easiest way of finding the cartesian equations of planes and lines in three dimensions.

Chapter 5 acts as an introduction to vector algebra by considering vector spaces, although the only vector spaces looked at in detail will be $\mathbb{R}^2$, $\mathbb{R}^3$ and some subspaces of these. This leads on, via linear dependence and independence, and the idea of a basis, to the discussion of transformation geometry in Chapter 6. Not many years ago, transformation geometry in two dimensions would have been met at GCSE level (or O-level, as it was then). Now this is no longer the case, it seems appropriate to include this topic in a book on geometric vectors, as, when one fully appreciates the two-dimensional case, the progression to three-dimensional transformation geometry is a small step. The basic geometrical results are expanded upon and linked with eigenvalues and eigenvectors, looked at again in a geometric light, and some useful special cases are considered.

The concept of *isometry* – something which preserves both shape and size – is central to the study of geometry, and this is developed in Chapter 7, partly as an exercise in using vectors and their scalar and vector products, and partly as a pointer to further studies in geometry. Vector calculus is introduced in Chapter 8 and Chapter 9 as a means of studying curves and surfaces in three-dimensional space, and it is vector calculus which paves the way to differential geometry and the world of manifolds and tangent bundles. These last three chapters are both a vehicle for consolidating the ideas found earlier in the book and a stepping stone to more advanced theories.

I would like to thank my colleague Ray d'Inverno for suggesting that I write this book, Professor Chris Collinson, who, as editor of the series, encouraged me from afar, and my colleagues and students who have given me so many new slants on the subject matter over the years. Thanks go to David Firth for introducing me to the delights of P$_I$CT$_E$X for preparing diagrams, and to many of my colleagues whose brains have been picked concerning the use of T$_E$X in which this text was prepared. I am particularly grateful to Susan Ward who read through the chapters of this book as they were written. Her careful attention to detail has helped to eradicate many of the host of errors that occurred in a first draft, and her view as a student on the receiving end was invaluable. That she found time to do this while studying full time for a degree and looking after her family is a tribute to her ability to organise her time so competently. Lastly, and most important of all, my thanks go to my husband, Keith, without whose constant encouragement and support (both moral and gastronomic) this book would not have been written.

1 • Introduction to Vectors

1.1 Vectors and scalars

When people ask 'What is a vector?' it is as difficult to answer as 'What is a number?' Both vectors and numbers are abstract ideas which represent more concrete quantities. We start by learning that two apples added to two apples gives us four apples, two pencils added to two pencils gives us four pencils, and so on using physical objects, and it is some time before we link this with the more abstract concept $2 + 2 = 4$. With a vector there are two quantities involved in the representation, and we generally think of these as *magnitude* and *direction*, and we often use the term *length* as an alternative for *magnitude*. So a vector is defined as something having both magnitude and direction, and anything which has just a magnitude attached to it is called a *scalar*. In this book all our scalars will be real numbers, but readers should be aware that there *are* vector spaces for which the scalars are complex numbers or even more exotic beings.

One way of differentiating between vectors and scalars is by considering the difference between the distance between two points, which is a scalar, and the displacement of one point from another, which is a vector, and which we can regard as what we have to do to get from one point to another. In this case we need to know not only *how far* we have to go, but also *in which direction*. Buckingham Palace is 1.25 km from Trafalgar Square, but if someone is starting from Trafalgar Square and wishes to get to Buckingham Palace, it is no good walking 1.25 km to the east!

Examples of scalar quantities are distances, speeds and masses, and examples of vector quantities are displacements, velocities, weights.

Notation

We shall represent vectors in bold type and scalars will be written in italics, so **v** represents a vector, but *s* represents a scalar. The vector which represents displacement from a point *A* to a point *B* will be written as **AB**, and the vector from *B* to *A* as **BA**. This indicates how vital it is to make the direction clear on a diagram, and we shall use arrows to indicate direction, as in Fig 1.1.

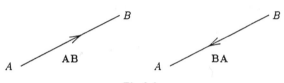

Fig 1.1

1.2 Basic definitions and notation

This book is concerned mainly with vectors in two or three dimensions. From a fairly early stage we are used to dealing in two dimensions by choosing two axes of coordinates, and thereby labelling every point by its own *coordinates*. A similar method is used in three dimensions. This, in essence, is what we do with vectors, whether in two or three (and possibly later on four, five, or even 42) dimensions. In two dimensions we generally label a point by the coordinates (x, y), and in three dimensions by (x, y, z), with respect to some given set of axes. Thus (1,2,3) will be the coordinates of the point whose displacement from the origin is 1 unit in the direction of the positive x-axis, 2 units in the direction of the positive y-axis, and 3 units in the direction of the positive z-axis. We call the set of real numbers $\mathbb{R}$, the two-dimensional plane $\mathbb{R}^2$ and three-dimensional space $\mathbb{R}^3$, indicating that two or three real number coordinates are needed in the respective spaces.

The Zero Vector, the Origin and Position Vectors

0 denotes the *zero vector* which is the only vector to which a direction cannot be assigned. If a man is standing at a particular place and takes a pace forwards then, if he is facing north, he will arrive at a different position from the place he would arrive at if he were facing east. However, if he took no paces at all, then he would be on the same spot no matter which direction he faced, and hence direction has no meaning in this particular case.

In order to get our bearings in any place, we need a base point to relate to. When visiting a new town or city we find a fixed display map with a *you are here* label extremely useful. We can choose any point in $\mathbb{R}^2$ or $\mathbb{R}^3$ to be an *origin* or base point, and once we have chosen this, all points are given in relation to this chosen origin, which we normally label O. The *position vector* of a point A is defined to be the displacement of that point from the origin O. With the notation discussed above, we can denote this by **OA**. We shall use a convention wherever possible that the position vector of a point A is denoted by **a**, the lower case version of the letter used to name the point. However, any vector with the same length and direction as the position vector **a** is regarded as being equal to **a**. If the vector does not have its tail tied to any particular point we call it a *free vector*.

Multiplication of a Vector by a Scalar

$-\mathbf{v}$ is a vector of the same magnitude or length as **v** but the opposite direction. With the notation above this means that **BA** $= -\mathbf{AB}$. We can multiply a vector **v** by a scalar α to give a vector $\alpha\mathbf{v}$ of length $|\alpha|$ times the length of **v**, in either the same direction as **v** if α is positive or the opposite direction to **v** if α is negative. Clearly, if α is zero, then $\alpha\mathbf{v}$ is the zero vector. This means that, if **a** and **b** are non-zero vectors, the vector **a** is *parallel* to the vector **b** if and only if **a** is a non-zero multiple of **b**.

EXERCISE 1

In Fig 1.2 we can see that $\mathbf{b} = 2\mathbf{a}$. Write down the other vectors shown in terms of **a**. The dotted lines are drawn to enable you to compare lengths.

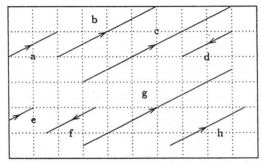

Fig 1.2

Suppose we choose an origin and a set of mutually orthogonal axes in $\mathbb{R}^3$, and label them as the *x*-axis, *y*-axis and *z*-axis in a right hand sense. This means that if, with a right hand, we put thumb, index finger and middle finger at right angles in the most natural way – as illustrated in Fig 1.3 – and let the thumb give the direction of the *x*-axis, the index finger the direction of the *y*-axis, and the middle finger the direction of the *z*-axis, then we have a right-handed system of axes.

Fig 1.3

EXERCISE 2

Which of the systems of axes in Fig 1.4 are right hand systems, and which are left hand systems? In each case the axis which is not drawn parallel to an edge of the page is regarded as pointing 'out of the page'.

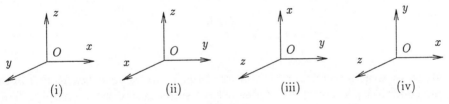

Fig 1.4

Unit Vectors

A *unit vector* is a vector whose length or magnitude is 1 unit. We can combine the two notions of unit vector and right-handed rectangular axes to form a basis for the vectors in $\mathbb{R}^3$. Suppose $\mathbf{i}, \mathbf{j}$ and $\mathbf{k}$ are unit vectors in the directions of the positive x-, y- and z-axes respectively; then since the axes form a right-handed system, we refer to the triad of vectors $\mathbf{i}, \mathbf{j}, \mathbf{k}$ as a right hand triad. Now the position vector of the point (x, y, z) can be expressed as

$$x\mathbf{i} + y\mathbf{j} + z\mathbf{k}.$$

In fact, because any vector $\mathbf{v}$ in $\mathbb{R}^3$ is equal to some position vector, *any* vector in $\mathbb{R}^3$ can be written in the form

$$p\mathbf{i} + q\mathbf{j} + r\mathbf{k}; \qquad \text{for some } p, q, r \in \mathbb{R}.$$

(The symbol $\in$ means 'in' or 'belonging to' and will be used throughout this book.)
   If we are working in $\mathbb{R}^2$, then any vector can be written in the form

$$p\mathbf{i} + q\mathbf{j}; \qquad \text{for some } p, q \in \mathbb{R}.$$

EXERCISE 3

Write down the position vectors of the points shown in Fig 1.5 in terms of $\mathbf{i}, \mathbf{j}$ and $\mathbf{k}$.

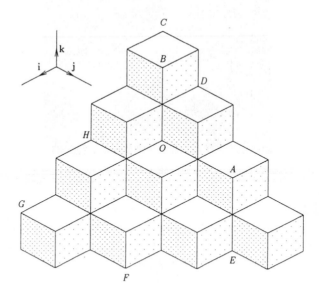

Fig 1.5

When we write a vector in this form we call a_1, a_2, a_3 the *components* of $\mathbf{a}$ in the directions of $\mathbf{i}, \mathbf{j}, \mathbf{k}$ respectively. Often, when the context is clear, we simply refer to a_1, a_2, a_3 as the *components* of $\mathbf{a}$.

Notation

We shall also use the notation

$$\begin{pmatrix} p \\ q \end{pmatrix}$$ to denote the vector $p\mathbf{i} + q\mathbf{j} \in \mathbb{R}^2$, and

$$\begin{pmatrix} p \\ q \\ r \end{pmatrix}$$ to denote the vector $p\mathbf{i} + q\mathbf{j} + r\mathbf{k} \in \mathbb{R}^3$,

and we shall refer to these representations on the left as *column vectors* for the obvious reason that they are written as columns.

EXERCISE 4

Write the vectors in the previous exercise as column vectors.

Length of a Vector

Let X be the point with coordinates (x, y, z) in $\mathbb{R}^3$, whose position vector $\mathbf{x}$ can be written in the form

$$\mathbf{x} = x\mathbf{i} + y\mathbf{j} + z\mathbf{k}.$$

Then the length of the vector is the distance of X from O, and by the three-dimensional version of Pythagoras' Theorem we know that this is

$$\sqrt{x^2 + y^2 + z^2}.$$

Notation

We use the notation $|\mathbf{v}|$ for the length of the vector $\mathbf{v}$.

Thus if A is the point $(3, -1, 2)$ and $\mathbf{a}$ is the position vector of A, we have

$$\mathbf{a} = 3\mathbf{i} - \mathbf{j} + 2\mathbf{k} \quad \text{and} \quad |\mathbf{a}| = \sqrt{3^2 + (-1)^2 + 2^2} = \sqrt{14}.$$

EXERCISE 5

Find the lengths of the following vectors

    (i) $\mathbf{i} + \mathbf{j} + \mathbf{k}$,   (ii) $2\mathbf{i} + \mathbf{j} - 2$,

    (iii) $2\mathbf{i} - 3\mathbf{j} + 6\mathbf{k}$,   (iv) $5\mathbf{i} - 3\mathbf{j} - 4\mathbf{k}$.

To find a unit vector in the direction of a known vector all we need to do is multiply the known vector by the reciprocal of its length. We shall use the notation $\hat{\mathbf{a}}$ for the unit vector in the direction of $\mathbf{a}$. So, in general

$$\hat{\mathbf{a}} = \frac{1}{|\mathbf{a}|}\mathbf{a}$$

and in the example above the vector **a** has length $\sqrt{14}$, so a unit vector in this direction is

$$\hat{\mathbf{a}} = \frac{1}{\sqrt{14}}(3\mathbf{i} - \mathbf{j} + 2\mathbf{k}).$$

If $\mathbf{a} = a_1\mathbf{i} + a_2\mathbf{j} + a_3\mathbf{k}$ we refer to the quantities

$$\frac{a_1}{|\mathbf{a}|}, \frac{a_2}{|\mathbf{a}|}, \frac{a_3}{|\mathbf{a}|}$$

as the *direction cosines* of **a**. Thus the direction cosines of **a** are the *components* of â. The reasons for this terminology will be given in Chapter 3.

1.3 Addition of vectors

We have seen how we multiply a vector by a real number. It is also possible to add vectors together, provided they are in the same space. Thus if our vectors **v**, **w** are in $\mathbb{R}^2$ with

$$\mathbf{v} = \begin{pmatrix} a \\ b \end{pmatrix}, \quad \mathbf{w} = \begin{pmatrix} p \\ q \end{pmatrix}, \quad \text{then} \quad \mathbf{v} + \mathbf{w} = \begin{pmatrix} a+p \\ b+q \end{pmatrix}.$$

Figure 1.6 illustrates why this works, for if $\mathbf{v} = \begin{pmatrix} 3 \\ 4 \end{pmatrix}$ and $\mathbf{w} = \begin{pmatrix} 2 \\ -3 \end{pmatrix}$, we see that the total effect of moving 3 units in the x-direction and 4 units in the y-direction, followed by 2 units in the x-direction and -3 units in the y-direction, is the same as moving 5 ($= 3 + 2$) units in the x-direction and 1 ($= 4 - 3$) unit in the y-direction.

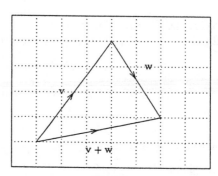

Fig 1.6

In a similar way if **v**, **w** are in $\mathbb{R}^3$ with

$$\mathbf{v} = \begin{pmatrix} a \\ b \\ c \end{pmatrix}, \quad \mathbf{w} = \begin{pmatrix} p \\ q \\ r \end{pmatrix}, \quad \text{then} \quad \mathbf{v} + \mathbf{w} = \begin{pmatrix} a+p \\ b+q \\ c+r \end{pmatrix}.$$

Parallelogram of Vectors

We can think of this in a more general way. If the sides AB and AD represent in both magnitude and direction the vectors **v** and **w** respectively, then if C is the point which makes $ABCD$ a parallelogram, the diagonal AC represents in both magnitude and direction the vector **v** + **w**, and since $a + p = p + a$ etc. (as a and p are real numbers), AC also represents **w** + **v** in magnitude and direction. That is to say

v + **w** = **w** + **v**

and we say that vector addition is *commutative*.

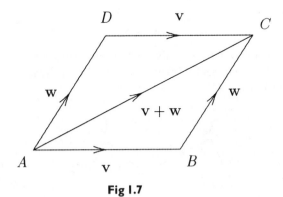

Fig I.7

Triangle of Vectors

This result leads to another, often more convenient, result. Because $ABCD$ is a parallelogram, the side BC represents the same magnitude and direction as does AD, so from above it follows that if the sides AB and BC of a triangle ABC represent the vectors **v** and **w** respectively in both magnitude and direction, then the third side of the triangle AC represents **v** + **w** in both magnitude and direction, and we represent this by

AB + **BC** = **AC**.

Both of the above results apply to vectors in $\mathbb{R}^3$ just as much as to $\mathbb{R}^2$.

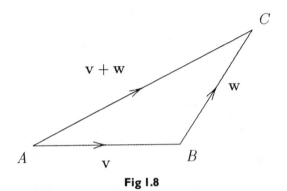

Fig I.8

EXERCISE 6

Given the vectors $\mathbf{v} = \begin{pmatrix} 1 \\ 2 \\ 3 \end{pmatrix}$ and $\mathbf{w} = \begin{pmatrix} 2 \\ -2 \\ 1 \end{pmatrix}$, find:

(i) $\mathbf{v} + \mathbf{w}$,   (ii) $2\mathbf{v} + 3\mathbf{w}$,   (iii) $4\mathbf{v} - 5\mathbf{w}$,   (iv) $\frac{1}{3}\mathbf{v} + \frac{2}{3}\mathbf{w}$.

If we extend this to more than two vectors, we can use the triangle of vectors result repeatedly to obtain the following. If $\mathbf{v}_1$, $\mathbf{v}_2$, ... , $\mathbf{v}_n$ are vectors which are represented in both magnitude and direction by the line segments $\mathbf{A}_1\mathbf{A}_2$, $\mathbf{A}_2\mathbf{A}_3$, ..., $\mathbf{A}_n\mathbf{A}_{n+1}$, then

$$\mathbf{v} = \mathbf{v}_1 + \mathbf{v}_2 + \ldots + \mathbf{v}_n = \mathbf{A}_1\mathbf{A}_2 + \mathbf{A}_2\mathbf{A}_3 + \ldots + \mathbf{A}_n\mathbf{A}_{n+1} = \mathbf{A}_1\mathbf{A}_{n+1}.$$

It should be noted that this only works if the vectors are *nose to tail*, that is if they follow one after the other following the direction of the arrows along a continuous path as shown in Fig 1.9.

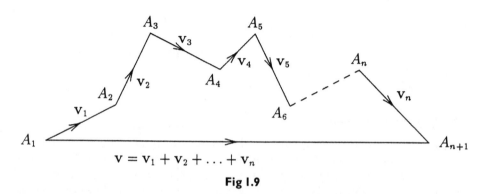

$$\mathbf{v} = \mathbf{v}_1 + \mathbf{v}_2 + \ldots + \mathbf{v}_n$$

Fig 1.9

In particular if we have three vectors $\mathbf{u}$, $\mathbf{v}$, $\mathbf{w}$ whose respective magnitudes and directions are represented by the line segments AB, BC, CA, then by using the triangle of vectors rule on triangle ABC we have

$$\mathbf{AB} + \mathbf{BC} = \mathbf{AC}$$

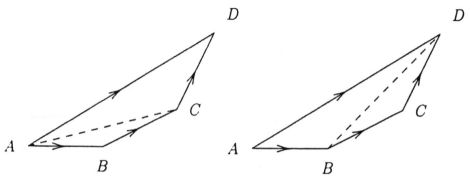

Fig 1.10

and then using the triangle of vectors rule on triangle ACD we have

$$\mathbf{AC} + \mathbf{CD} = \mathbf{AD}.$$

We could also use the triangle of vectors rule first on the triangle BCD to get

$$\mathbf{BC} + \mathbf{CD} = \mathbf{BD}$$

and then on triangle ABD to get

$$\mathbf{AB} + \mathbf{BD} = \mathbf{AD}.$$

From the two expressions for **AD** we see that

$$(\mathbf{AB} + \mathbf{BC}) + \mathbf{CD} = \mathbf{AB} + (\mathbf{BC} + \mathbf{CD})$$

or

$$(\mathbf{u} + \mathbf{v}) + \mathbf{w} = \mathbf{u} + (\mathbf{v} + \mathbf{w}).$$

This tells us that vector addition is *associative*.

Summary

1. Definitions for vectors, scalars, zero vector, position vectors, free vectors and multiplication of a vector by a scalar were given.
2. We considered right hand systems of axes, the length of a vector, unit vectors, the vectors **i**, **j**, **k**, components and direction cosines.
3. The triangle of vectors and the parallelogram of vectors show that

$$\mathbf{u} + \mathbf{v} = \mathbf{v} + \mathbf{u} \quad \text{and} \quad \mathbf{u} + (\mathbf{v} + \mathbf{w}) = (\mathbf{u} + \mathbf{v}) + \mathbf{w}.$$

This means that addition of vectors is *commutative* and *associative*.

FURTHER EXERCISES

7. Let A be the point in $\mathbb{R}^2$ whose coordinates are $(1, 2)$, and whose position vector is **a**. On squared paper draw the vectors **a**, **2a**, **−a**, **−3a**, $\frac{1}{2}$**a**.
8. Repeat Exercise 7 with A having coordinates $(2, -3)$.
9. Collect into sets of parallel vectors the following:

$$\mathbf{c} = \begin{pmatrix} 1 \\ 2 \\ 3 \end{pmatrix}, \quad \mathbf{d} = \begin{pmatrix} 3 \\ 2 \\ 1 \end{pmatrix}, \quad \mathbf{e} = \begin{pmatrix} -1 \\ 1/2 \\ -1/2 \end{pmatrix}, \quad \mathbf{f} = \begin{pmatrix} 1/6 \\ 1/3 \\ 1/2 \end{pmatrix}, \quad \mathbf{g} = \begin{pmatrix} 1/2 \\ -1/4 \\ 1/4 \end{pmatrix}.$$

10. Find the lengths of the vectors given in Exercise 9.
11. Let $\mathbf{a} = 2\mathbf{i} - \mathbf{j} + 2\mathbf{k}$ and $\mathbf{b} = 6\mathbf{i} + 2\mathbf{j} - 3\mathbf{k}$. Write down in terms of **i**, **j** and **k**, the following vectors:

$$\mathbf{c} = \mathbf{a} + \mathbf{b}, \quad \mathbf{d} = \mathbf{b} - \mathbf{a}, \quad \mathbf{e} = \mathbf{a} + \mathbf{b} + \mathbf{c} + \mathbf{d}, \quad \mathbf{f} = 2\mathbf{a} + 3\mathbf{b}, \quad \mathbf{g} = 2\mathbf{b} - 6\mathbf{a}.$$

Illustrate these sums on a diagram similar to Fig 1.7 or 1.8.
12. Find unit vectors in the directions of the vectors **a**, **b**, **c**, **d**, **e**, **f** and **g** of Exercise 11.

2 • Vector Equation of a Straight Line

2.1 The vector equation of a straight line

The equation of a straight line in a plane will be familiar to students before embarking upon a course involving vectors. The general equation of a line in the plane $\mathbb{R}^2$ is

$$ax + by + c = 0.$$

So is there a similar equation for a line in $\mathbb{R}^3$? Suppose we look at the equation

$$ax + by + cz + d = 0.$$

This does not define a line, but, as we shall see later, a plane in $\mathbb{R}^3$. In fact it is much easier to find the *vector* equation of a line in $\mathbb{R}^3$ than it is to find the *cartesian* equation.

There are two useful versions of the vector equation of a line, and the one we choose depends upon what information is given.

Case (i) To Find the Vector Equation of a Line l Given a Fixed Point on the Line and a Vector Parallel to the Line l

Suppose the fixed point on the line is A whose position vector with respect to our chosen origin is $\mathbf{a}$, and the vector parallel to the line is $\mathbf{c}$. Let R be any point on l, and let the position vector of R be $\mathbf{r}$. This configuration is illustrated in Fig 2.1.

From the diagram we can see that if we move from O to R directly it has the same overall effect as moving from O to A and then from A to R. As a result of the addition of vectors which we met in the previous chapter, we can say that

$$\mathbf{OR} = \mathbf{OA} + \mathbf{AR}, \tag{2.1.1}$$

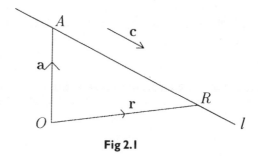

Fig 2.1

but we know that **AR** is parallel to the vector **c**, and therefore **AR** $= \lambda$**c** for some real number λ. So now we can rewrite (2.1.1) as

$$\mathbf{r} = \mathbf{a} + \lambda\mathbf{c}, \tag{2.1.2}$$

and this is the vector equation of the line *l* which is parallel to **c** and passes through the point *A*.

Case (ii) To Find the Vector Equation of a Line *l* When Two Points on the Line are Given

Suppose the two given points are *A* and *B* whose position vectors are respectively **a** and **b**. This configuration is illustrated in Fig 2.2.

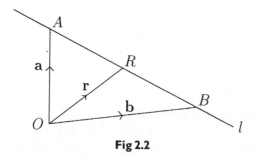

Fig 2.2

We can see from the diagram that since *AB* lies along the line *l*, the vector **AB** is parallel to *l*. Now displacement from *A* to *B* is equivalent to displacement from *A* to *B* via any intermediate point, and in particular through *O*. So in vector terms

$$\mathbf{AB} = \mathbf{AO} + \mathbf{OB}. \tag{2.1.3}$$

We already know that **OA** $=$ **a**, and that **OB** $=$ **b**, so that **AO** $=$ $-$**a** from our observations in Chapter 1. Thus (2.1.3) can be rewritten as

$$\mathbf{AB} = -\mathbf{a} + \mathbf{b} = \mathbf{b} - \mathbf{a}. \tag{2.1.4}$$

This means that we now know a point on the line *l* and a vector parallel to *l*, so substituting into (2.1.2) we get

$$\mathbf{r} = \mathbf{a} + \lambda(\mathbf{b} - \mathbf{a}).$$

Tidying this up a little, we can write it in the form

$$\mathbf{r} = (1 - \lambda)\mathbf{a} + \lambda\mathbf{b} \tag{2.1.5}$$

and this is the required vector equation for the line *l* which passes through the points *A* and *B*.

2.2 The cartesian equations of a straight line

The vector equation (2.1.2) is the key to finding the cartesian equations of a straight line *l*, for if we write vectors in terms of the three unit vectors **i**, **j** and **k** (suppose **a** $= a_1\mathbf{i} + a_2\mathbf{j} + a_3\mathbf{k}$, **c** $= c_1\mathbf{i} + c_2\mathbf{j} + c_3\mathbf{k}$ and **r** $= x\mathbf{i} + y\mathbf{j} + z\mathbf{k}$), the equation (2.1.2) could be written in coordinate form as

$$x = a_1 + \lambda c_1, \quad y = a_2 + \lambda c_2, \quad z = a_3 + \lambda c_3$$

which we could convert into

$$\frac{x - a_1}{c_1} = \frac{y - a_2}{c_2} = \frac{z - a_3}{c_3} = \lambda.$$

Thus if we are working in cartesian coordinates, the equations of the line l which passes through the point (a, b, c) and is parallel to the line joining $(0, 0, 0)$ to (p, q, r) are

$$\frac{x - a}{p} = \frac{y - b}{q} = \frac{z - c}{r}. \tag{2.2.1}$$

We need to take into account the special cases where one or two of p, q, r are zero. If $p = 0$, say, then our equations become

$$x = a, \quad \frac{y - b}{q} = \frac{z - c}{r}$$

and our straight line is a line in the plane $x = a$, whilst if $p = q = 0$ we can define our line by $x = a$, $y = b$, and we get a line parallel to the z-axis. Clearly, we cannot have the situation where $p = q = r = 0$, for then $\mathbf{c}$ would be the zero vector, and no direction for the line would be defined.

● *Example 1*

Find the vector equation of the line passing through $A = (1, 2, 3)$ and $B = (-2, 1, -3)$. Which of the following points lie on the line: $(1, -1, 4)$, $(4, 3, 9)$? Find also the cartesian equations of the line.

SOLUTION
From (2.1.5) the vector equation of the line is

$$\mathbf{r} = (1 - \lambda)\begin{pmatrix} 1 \\ 2 \\ 3 \end{pmatrix} + \lambda\begin{pmatrix} -2 \\ 1 \\ -3 \end{pmatrix} = \begin{pmatrix} 1 - \lambda - 2\lambda \\ 2 - 2\lambda + \lambda \\ 3 - 3\lambda - 3\lambda \end{pmatrix} = \begin{pmatrix} 1 - 3\lambda \\ 2 - \lambda \\ 3 - 6\lambda \end{pmatrix}.$$

So,

$$\begin{pmatrix} x \\ y \\ z \end{pmatrix} = \begin{pmatrix} 1 - 3\lambda \\ 2 - \lambda \\ 3 - 6\lambda \end{pmatrix}. \tag{2.2.2}$$

Now, $(1, -1, 4)$ lies on the line if and only if we can find a value of λ which satisfies all three equations

$$1 - 3\lambda = 1, \quad 2 - \lambda = -1, \quad 3 - 6\lambda = 4.$$

However, the first equation implies that $\lambda = 0$, but this does not satisfy the other two, and hence no value of λ will satisfy all three equations, so $(1, -1, 4)$ does not lie on the line.

Secondly,

$$1 - 3\lambda = 4, \quad 2 - \lambda = 3, \quad 3 - 6\lambda = 9$$

are all satisfied by $\lambda = -1$; therefore the point $(4, 3, 9)$ does lie on the line l.

Going back to equation (2.2.2) we see that the equations of the line could be written as

$$\frac{x-1}{3} = -\lambda, \quad \frac{y-2}{1} = -\lambda, \quad \frac{z-3}{6} = -\lambda$$

or $\quad \dfrac{x-1}{3} = \dfrac{y-2}{1} = \dfrac{z-3}{6} \quad (= -\lambda)$

which is the required set of cartesian equations of the line.

By choosing different values of λ we can find different points on this line. For example, $\lambda = 1$ gives us the point $(-2, 1, -3)$, and $\lambda = -2$ gives us the point $(7, 4, 15)$.

EXERCISE I

Find both the vector equation and the cartesian equations of the line passing through the points A $(1, 2, -3)$ and B $(-1, 1, 2)$. Write down the coordinates of three other points on the line.

2.3 A point dividing a line segment in a given ratio

An equation which yields more than is first apparent is (2.1.5). Consider a point P which lies on the line AB a quarter of the way along from A. Then $AP = \frac{1}{4}AB$, and in terms of vectors, $\mathbf{AP} = \frac{1}{4}\mathbf{AB}$.

Then, as usual, we have

$$\mathbf{OP} = \mathbf{OA} + \mathbf{AP}$$

$$= \mathbf{OA} + \frac{1}{4}\mathbf{AB}$$

$$= \mathbf{a} + \frac{1}{4}(\mathbf{b} - \mathbf{a})$$

and hence,

$$\mathbf{p} = \frac{3}{4}\mathbf{a} + \frac{1}{4}\mathbf{b}. \tag{2.3.1}$$

If we compare this with equation (2.1.5), we see that the right hand side of equation (2.3.1) is the same as the right hand side of equation (2.1.5) with λ replaced by $\frac{1}{4}$. But this is not really surprising since if $AP = \lambda AB$, whatever the value of λ we will get the expression given in equation (2.1.5) by following the working through in exactly the same way.

In particular, if M is the *midpoint* of AB, since $\lambda = (1 - \lambda) = 1/2$, the position vector of M is

$$\mathbf{m} = \frac{1}{2}(\mathbf{a} + \mathbf{b}). \tag{2.3.2}$$

At first sight it seems as though the fractions in (2.3.1) have got themselves the wrong way round since $AP : PB = \frac{1}{4} : \frac{3}{4}$, but if we remember that in this case P is closer to A than it is to B, then it is reasonable to suppose that $\mathbf{a}$ will have a greater part to play than $\mathbf{b}$ in the expression of $\mathbf{p}$. It is always useful to test whether our answer is reasonable in this branch of mathematics as well as any other, and in this case we see that our findings *are* reasonable.

From the discussion above it becomes clear that P divides the line segment AB in the ratio $\lambda : (1 - \lambda)$, if and only if the position vector of P is given by

$$\mathbf{p} = (1 - \lambda)\mathbf{a} + \lambda\mathbf{b}. \tag{2.3.3}$$

EXERCISE 2

On graph paper plot the points A and B whose position vectors are

$$\mathbf{a} = \begin{pmatrix} 2 \\ 0 \end{pmatrix} \quad \text{and} \quad \mathbf{b} = \begin{pmatrix} 0 \\ 3 \end{pmatrix}$$

respectively. Then for $\lambda = 0, 1, 2, -1, 1/2, 3/2, -1/2$ plot the points $(1 - \lambda)\mathbf{a} + \lambda\mathbf{b}$, and calculate $\lambda/(1 - \lambda)$, when $\lambda \neq 1$.

It will become apparent from this exercise that if $\lambda > 1$ or if $\lambda < 0$ the point P lies outside the line segment joining A and B and the ratio $AP/PB = \lambda/(1 - \lambda)$ (which is equivalent to $\lambda : (1 - \lambda)$) is negative, whereas if $0 < \lambda < 1$ the point P lies between A and B, and the ratio is positive. This means that when the *vectors* $\mathbf{AP}$ and $\mathbf{PB}$ point in the same direction, the ratio AP/PB is positive, but when they point in opposite directions, the ratio is negative. What happens when $\lambda = 0, 1$?

Unfortunately, we are not always given the precise fraction of the line segment AB which gives AP, but we may be told that P divides AB in the ratio $m : n$ where $m + n \neq 1$. In this case we need to be able to find a λ such that $m : n = \lambda : (1 - \lambda)$. Now

$$\frac{m}{n} = \frac{\lambda}{(1 - \lambda)} \iff (1 - \lambda)m = \lambda n \iff \lambda = \frac{m}{m + n}.$$

It then follows that

$$(1 - \lambda) = 1 - \frac{m}{m + n} = \frac{n}{m + n}.$$

This means that, having found λ we can now substitute into (2.1.5) to get

$$\mathbf{p} = \frac{n}{m + n}\mathbf{a} + \frac{m}{m + n}\mathbf{b}. \tag{2.3.4}$$

EXERCISE 3

(i) If the coordinates of A and B are $(-1, 6, -7)$ and $(4, -4, 3)$ respectively, find the position vector of the point P which divides AB in the ratio $3 : 2$.

(ii) In what ratio does Q divide the line segment BA if Q has coordinates $(5, -6, 5)$?

2.4 Points of intersection of lines

If two lines intersect at a point, then the position vector of the point must satisfy the vector equations of both lines. This is best illustrated with examples.

● *Example 2*

Find, if possible, the intersection of the following lines:

(i) the line through $(4, 5, 1)$ parallel to the vector $\begin{pmatrix} 1 \\ 1 \\ 1 \end{pmatrix}$ and the line through the

point $(5, -4, 0)$ parallel to the vector $\begin{pmatrix} 2 \\ -3 \\ 1 \end{pmatrix}$,

(ii) the line passing through the two points $(-1, 1, 2)$ and $(0, -2, 4)$, and the line passing through the two points $(1, 1, 3)$ and $(3, 0, 4)$.

SOLUTION

(i) Let us call the lines l and m respectively. Then any point on l has position vector

$$\mathbf{r} = \begin{pmatrix} 4 \\ 5 \\ 1 \end{pmatrix} + \lambda \begin{pmatrix} 1 \\ 1 \\ 1 \end{pmatrix} = \begin{pmatrix} 4 + \lambda \\ 5 + \lambda \\ 1 + \lambda \end{pmatrix}.$$

Similarly, any point on m has position vector

$$\mathbf{r} = \begin{pmatrix} 5 \\ -4 \\ 0 \end{pmatrix} + \mu \begin{pmatrix} 2 \\ -3 \\ 1 \end{pmatrix} = \begin{pmatrix} 5 + 2\mu \\ -4 - 3\mu \\ \mu \end{pmatrix}.$$

These lines intersect if and only if there is a point which is common to both lines. In this case the common point must have a position vector which can be expressed in both of the above ways. If the two different expressions represent the same position vector, then the coordinates must be identical. So we need to be able to find values of λ and μ which give the same position vector $\mathbf{r}$ in the two expressions above, and hence we need to solve

$$4 + \lambda = 5 + 2\mu \tag{2.4.1}$$

$$5 + \lambda = -4 - 3\mu \tag{2.4.2}$$

$$1 + \lambda = \mu. \tag{2.4.3}$$

From (2.4.3) substitute in (2.4.1) to get

$$4 + \lambda = 5 + 2(1 + \lambda)$$

which gives the solution $\lambda = -3$ and $\mu = -2$. This also satisfies equation (2.4.2) and hence it satisfies all three equations. Substituting these values in the equations

for $\mathbf{r}$ in each case gives $\mathbf{r} = \begin{pmatrix} 1 \\ 2 \\ -2 \end{pmatrix}$, and this is the position vector of the point of

intersection of the two lines.

(ii) In this case we have to solve

$$(1 - \lambda)\begin{pmatrix} -1 \\ 1 \\ 2 \end{pmatrix} + \lambda \begin{pmatrix} 0 \\ -2 \\ 4 \end{pmatrix} = (1 - \mu)\begin{pmatrix} 1 \\ 1 \\ 3 \end{pmatrix} + \mu \begin{pmatrix} 3 \\ 0 \\ 4 \end{pmatrix}.$$

This can be written in the form

$$-1 + \lambda = 1 + 2\mu \tag{2.4.4}$$

$$1 - 3\lambda = 1 - \mu \tag{2.4.5}$$

$$2 + 2\lambda = 3 + \mu. \tag{2.4.6}$$

From (2.4.5) $\mu = 3\lambda$, and substituting this into (2.4.4) gives $\lambda = -2/5$ and $\mu = -6/5$; however, upon substitution into (2.4.6) we see that the LHS = 6/5, but the RHS = 9/5. Therefore there is no solution to the set of three equations (2.4.4)–(2.4.6), and hence the two lines do not intersect.

There are two possible cases where two lines do not intersect. In one case they are *parallel*, and in the other case they are *skew*. These cases can be illustrated by the edges of a rectangular box.

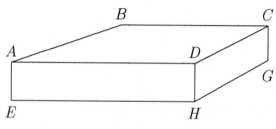

Fig 2.3

Consider here the lines which pass along the edges of the box. We still think of lines as being infinitely extended in both directions. Then the lines which pass along edges EH and GH clearly meet at H. However, the lines AD and EH do not meet – they are parallel – and the lines BD and EG do not meet – they lie in parallel planes – and such lines, if they are not parallel, are called *skew* lines. So any two lines in $\mathbb{R}^3$ either meet in a single point, or they are parallel or skew. (We shall see later why, if two lines in $\mathbb{R}^3$ do not meet, they must lie in parallel planes.)

● *Example 3*

Suppose P and Q are points on the sides OA and OB respectively of a triangle OAB such that $OP = \frac{1}{2}OA$ and $OQ = \frac{1}{3}OB$. Let X be the intersection of the lines AQ and BP. If $\mathbf{a}$ and $\mathbf{b}$ are the position vectors of A and B with respect to the origin O, find the position vector of X with respect to O. Also, if Y is the point where OX cuts AB, find the ratio in which Y cuts AB.

SOLUTION

The first thing we need here is a diagram. When sketching a diagram for a solution, it does not have to be to scale, but if you make it *roughly* in the right proportion, it will tell you if your answer is reasonable or wildly incorrect.

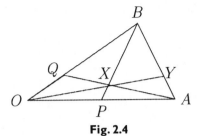

Fig. 2.4

Continuing with the convention that the bold lower case letter **p** represents the position vector of the point P etc., we know that

$$\mathbf{p} = \frac{1}{2}\mathbf{a} \quad \text{and} \quad \mathbf{q} = \frac{1}{3}\mathbf{b}.$$

Now since X lies on both AQ and BP we have

$$\mathbf{x} = (1 - \lambda)\mathbf{a} + \lambda\mathbf{q} \tag{2.4.7}$$

$$\mathbf{x} = (1 - \mu)\mathbf{b} + \mu\mathbf{p}. \tag{2.4.8}$$

From above we could rewrite these as

$$\mathbf{x} = (1 - \lambda)\mathbf{a} + \frac{1}{3}\lambda\mathbf{b} \quad \text{and} \quad \mathbf{x} = (1 - \mu)\mathbf{b} + \frac{1}{2}\mu\mathbf{a}.$$

This leads to

$$(1 - \lambda)\mathbf{a} + \frac{1}{3}\lambda\mathbf{b} = (1 - \mu)\mathbf{b} + \frac{1}{2}\mu\mathbf{a}$$

which in turn leads to

$$\left((1 - \lambda) - \frac{1}{2}\mu\right)\mathbf{a} = \left((1 - \mu) - \frac{1}{3}\lambda\right)\mathbf{b}.$$

Now if we assume that OAB is a non-degenerate triangle, then **a** cannot be parallel to **b** and the only possible way that the last equation could be true is if the coefficients of **a** and **b** are zero. That is, if

$$(1 - \lambda) - \frac{1}{2}\mu = 0 = (1 - \mu) - \frac{1}{3}\lambda.$$

Some fairly straightforward algebra will give the solution

$$\lambda = \frac{3}{5}, \qquad \mu = \frac{4}{5}.$$

Substituting these values into (2.4.7) and (2.4.8) will give the same value

$$\mathbf{x} = \frac{2}{5}\mathbf{a} + \frac{1}{5}\mathbf{b}$$

which is yet another check that we have the right answer here.

Now we need to find the position vector of the point Y. Since Y lies on OX we know that

$$\mathbf{y} = \alpha\mathbf{x} = \alpha\left(\frac{2}{5}\mathbf{a} + \frac{1}{5}\mathbf{b}\right)$$

for some α, and since Y lies on AB we have

$$\mathbf{y} = (1 - \beta)\mathbf{a} + \beta\mathbf{b}.$$

This can only be true if

$$1 - \beta = \frac{2}{5}\alpha \quad \text{and} \quad \beta = \frac{1}{5}\alpha.$$

This would give the solution $\alpha = 5/3$ and $\beta = 1/3$. A quick check tells us that these values give the same value for $\mathbf{y}$ from both equations. So, $\mathbf{y} = \frac{2}{3}\mathbf{a} + \frac{1}{3}\mathbf{b}$. Hence Y divides AB in the ratio $1:2$.

EXERCISE 4

The points A to F have coordinates as follows: $A\ (-4, 3, 2)$, $B\ (1, 1, -3)$, $C\ (-1, 2, 1)$, $D\ (-2, 4, 3)$, $E\ (2, 3, 0)$, $F\ (1, 4, 3)$. Show that the lines AD, BE, CF lie along the edges of a tetrahedron whose vertices are A, B, C and V, and find the coordinates of V. Find also the ratios in which D, E and F divide the line segments AV, BV and CV respectively. (*Hint*: Show that AD and BE intersect in a point V, and show also that V lies on CF.)

2.5 Some applications

The Median Centre or Centroid of a Triangle

This is the intersection of the three medians of the triangle. A median of a triangle is the line segment joining a vertex to the midpoint of the side opposite that vertex. But how do we show that these lines actually meet at a point? Once again, vectors gives us a straightforward way of calculating this.

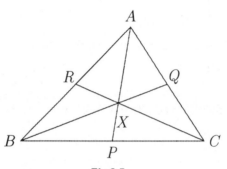

Fig 2.5

SOLUTION

We know from (2.3.2) that if the points A, B, C have respective position vectors **a**, **b**, **c**, then the respective midpoints P, Q, R of BC, CA, AB have respective position vectors $\frac{1}{2}(\mathbf{b} + \mathbf{c})$, $\frac{1}{2}(\mathbf{c} + \mathbf{a})$, $\frac{1}{2}(\mathbf{a} + \mathbf{b})$.

So if the three lines AP, BQ, CR are concurrent, there is a point which lies on all three lines. Now X lies on all three lines if and only if λ, μ and ν can be found so that its position vector **x** can be expressed in each of the following three ways:

$$\mathbf{x} = (1 - \lambda)\mathbf{a} + \lambda\mathbf{p}$$
$$\mathbf{x} = (1 - \mu)\mathbf{b} + \mu\mathbf{q}$$
$$\mathbf{x} = (1 - \nu)\mathbf{c} + \nu\mathbf{r}$$

but substituting for **p**, **q**, **r** in terms of **a**, **b**, **c** these three equations become

$$\mathbf{x} = (1 - \lambda)\mathbf{a} + \frac{1}{2}\lambda\mathbf{b} + \frac{1}{2}\lambda\mathbf{c}$$

$$\mathbf{x} = \frac{1}{2}\mu\mathbf{a} + (1 - \mu)\mathbf{b} + \frac{1}{2}\mu\mathbf{c}$$

$$\mathbf{x} = \frac{1}{2}\nu\mathbf{a} + \frac{1}{2}\nu\mathbf{b} + (1 - \nu)\mathbf{c}.$$

We can see that if $\lambda = \mu = \nu = 2/3$ then all three equations give the same vector value for **x**, and so the position vector for the centroid of triangle ABC is

$$\mathbf{x} = \frac{1}{3}(\mathbf{a} + \mathbf{b} + \mathbf{c}).$$

The fact that $\lambda = \mu = \nu = 2/3$ shows also that the centroid of the triangle divides each of the medians in the ratio $2{:}1$.

● *Menelaus' Theorem* (Menelaus of Alexandria c.100) ——

This is a very famous theorem in geometry, and proving it using coordinate geometry is very messy. However, using vectors its proof is much easier. The theorem says that if a straight line cuts the sides BC, CA, AB (produced if necessary) of a triangle ABC in the points P, Q, R respectively, then

$$\frac{BP}{PC} \cdot \frac{CQ}{QA} \cdot \frac{AR}{RB} = -1.$$

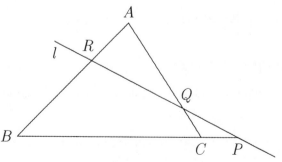

Fig 2.6

There are two possibilities here. As you found in Exercise 2, a point dividing the line segment in a negative ratio must lie outside that line segment. Now any line cutting one side of a triangle internally must cut one of the other two sides internally. Since it goes *into* the triangle it must also come out. (We can set aside the case where the line passes through a vertex.) It must therefore cut the third side externally. The only other possibility is that the line does not cut any side of the triangle internally. In each case there is an odd number of negative ratios, and therefore the product will be negative, so we can see that the negative sign will always be correct. We now need to look at the numbers.

Let us adopt the usual convention that **a** is the position vector of A, and so on. Then

$$\mathbf{p} = (1 - \lambda)\mathbf{b} + \lambda\mathbf{c}, \qquad \mathbf{q} = (1 - \mu)\mathbf{c} + \mu\mathbf{a}$$

for some real numbers λ, μ. Since R lies on the line PQ we have

$$
\begin{aligned}
\mathbf{r} &= (1 - \alpha)\mathbf{p} + \alpha\mathbf{q} \\
&= (1 - \alpha)\{(1 - \lambda)\mathbf{b} + \lambda\mathbf{c}\} + \alpha\{(1 - \mu)\mathbf{c} + \mu\mathbf{a}\} \\
&= \alpha\mu\mathbf{a} + (1 - \alpha)(1 - \lambda)\mathbf{b} + \{(1 - \alpha)\lambda + \alpha(1 - \mu)\}\mathbf{c}.
\end{aligned}
$$

But we also know that R lies on AB, so that

$$\mathbf{r} = (1 - \nu)\mathbf{a} + \nu\mathbf{b}.$$

Thus if

$$1 - \nu = \alpha\mu, \qquad \nu = (1 - \alpha)(1 - \lambda), \qquad \lambda + \alpha - \alpha\lambda - \alpha\mu = 0$$

both conditions are satisfied.

Now from the third of these equations we get $\alpha = \lambda/(\mu + \lambda - 1)$, and hence $(1 - \alpha) = (\mu - 1)/(\mu + \lambda - 1)$. Substituting this into the second equation gives $\nu = (\mu - 1)(1 - \lambda)/(\mu + \lambda - 1)$, and the first equation yields $1 - \nu = \mu\lambda/(\mu + \lambda - 1)$. This means that

$$\frac{1 - \nu}{\nu} = \frac{\mu\lambda}{(\mu - 1)(1 - \lambda)}.$$

Since $BP/PC = \lambda/(1 - \lambda)$ etc., we now have

$$\frac{BP}{PC} \cdot \frac{CQ}{QA} \cdot \frac{AR}{RB} = \frac{\lambda}{(1 - \lambda)} \cdot \frac{\mu}{(1 - \mu)} \cdot \frac{(\mu - 1)(1 - \lambda)}{\mu\lambda} = -1$$

and we have proved the theorem.

Summary

1. The vector equation of a line through A, parallel to the vector **c**, is

$$\mathbf{r} = \mathbf{a} + \lambda\mathbf{c}.$$

2. The vector equation of a line through the points A and B is

$$\mathbf{r} = (1 - \lambda)\mathbf{a} + \lambda\mathbf{b}.$$

3. The set of cartesian equations of the line through the point (a, b, c) and parallel to the line joining the origin to (p, q, r) is

$$\frac{x - a}{p} = \frac{y - b}{q} = \frac{z - c}{r} \qquad p, q, r \neq 0.$$

(Special cases where one or two of p, q, r are zero.)

4. The point dividing AB in the ratio $m : n$ has position vector

$$\left(\frac{n}{m + n}\right)\mathbf{a} + \left(\frac{m}{m + n}\right)\mathbf{b}.$$

5. We can find the point of intersection of two lines by comparing the coefficients of fixed vectors in the vector equations of the two lines. (A more rigorous approach will be seen later.)

FURTHER EXERCISES

5. Find both the vector equation and the cartesian equations of:
   (i) the line passing through the point $(2, 1, -2)$ parallel to the vector

$$\mathbf{c} = \begin{pmatrix} 4 \\ -2 \\ 3 \end{pmatrix},$$

   (ii) the line passing through the two points $(3, -1, 2)$ and $(1, 2, 4)$.
6. If A, B, C, D are the points whose respective coordinates are $(1, -1, 0)$, $(3, 1, 2)$, $(2, -2, 4)$, $(-1, 1, -1)$, which of the following points lie on AB, which lie on CD and which lie on neither: $(-2, 2, 4)$, $(2, -2, 4)$, $(-2, -2, -4)$, $(2, 0, 1)$, $(1, 3, -2)$, $(0, 2, 1)$, $(0, -2, -1)$?
7. Find, if possible, the point of intersection of the two lines in each of the following cases:

   (i) the line through $(5, 1, -3)$ parallel to the vector $\begin{pmatrix} 2 \\ 1 \\ -2 \end{pmatrix}$, and the line through $(2, -3, 0)$ parallel to the vector $\begin{pmatrix} -1 \\ 2 \\ 1 \end{pmatrix}$,

   (ii) the line through $(4, 1, 4)$ and $(3, -1, -6)$, and the line through $(-3, 0, -1)$ and $(2, 1, 4)$,

   (iii) the line through $(0, 0, 1)$ and $(2, 3, 5)$, and the line through $(0, 0, -1)$ and $(1, -2, 3)$.
8. Let A, B, C, D be any four points. Let P, Q, R, S be the midpoints of the line segments AB, BC, CD, DA respectively. With the usual convention, find $\mathbf{p}, \mathbf{q}, \mathbf{r}, \mathbf{s}$ in terms of $\mathbf{a}, \mathbf{b}, \mathbf{c}, \mathbf{d}$, and show that $PQRS$ is a parallelogram. (*Hint:* Find the vectors $\mathbf{PQ}$ and $\mathbf{SR}$ in terms of $\mathbf{a}, \mathbf{b}, \mathbf{c}, \mathbf{d}$.)
9. A spire of a church is in the shape of a right pyramid with vertex V, and a square base $ABCD$. A lightning conductor is to be put down the edge VC. Suppose a coordinate system is chosen so that the origin is in the plane of $ABCD$, and A, B, C, D lie at $(0, 0, 0)$, $(3, 0, 0)$, $(3, 3, 0)$, $(0, 3, 0)$ respectively, where the units represent the number of metres. Suppose also that the edge AV of the spire is in the direction of the vector $(1, 1, 4)$.

(i) Find the position vector of V, and the length of the edge along which the lightning conductor is fixed.

(ii) Find the cartesian equations of the four edges AV, BV, CV, DV.

(iii) If a man is standing on the ground with his feet at the point Q with position vector $(8, 15, -20)$, what is the distance from his feet to the top of the tower, and the vector equation of the line joining the points V and Q? If the man is 2 m tall, how high above his head is the top of the spire?

10. **Ceva's Theorem** (Giovanni Ceva 1648–1734)

    If P, Q, R lie on the respective sides BC, CA, AB of the triangle ABC, and if AP, BQ, CR are concurrent (that is, they all meet at a point) then

    $$\frac{BP}{PC} \cdot \frac{CQ}{QA} \cdot \frac{AR}{RB} = 1.$$

    Prove this theorem. The proof of Menelaus' Theorem will help you to get started. Let the point of intersection of the three lines be X, and find $\mathbf{x}$ in terms of $\mathbf{a}, \mathbf{b}, \mathbf{c}$ in three different ways (from the vector equations of the three lines). Equate the respective coefficients of $\mathbf{a}, \mathbf{b}, \mathbf{c}$ and this should lead you to the required result.

11. **CHALLENGE QUESTION** Suppose $ABCDEF$ is a hexagon whose opposite sides are parallel, but not necessarily equal in length. Show that the lines joining midpoints of opposite sides are concurrent.

12. **Project** Investigate the Example 3 given in Section 2.4. If in our example $OP = \frac{1}{4}OA$ and $OQ = \frac{1}{7}OB$, find the position vector of X and the ratio in which Y divides the line AB. Can you find a general result? Can you prove that this result works in all cases? Are there any exceptions?

3 • Scalar Products and Equations of Planes

3.1 The scalar product

We can add vectors together, and we can multiply a vector by a scalar, but so far we have not 'multiplied' vectors. In fact there are two kinds of vector 'product', and in both cases the element of multiplication is only part of the story.

● Definition I

The *scalar product* of two vectors **a** and **b** is defined to be

$$\mathbf{a}.\mathbf{b} = |\mathbf{a}||\mathbf{b}| \cos \theta,$$

where θ is the angle between the directions of the vectors **a** and **b**.

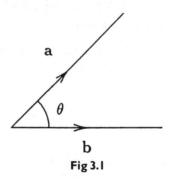

Fig 3.1

The name *scalar product* refers to the fact that the result of a scalar product is *scalar*. It is sometimes called the 'dot product', because of the dot in the expression which should always be clearly shown to tell us what sort of product this is.

Length of a Vector

Since $\cos 0 = 1$,

$$\mathbf{a}.\mathbf{a} = |\mathbf{a}||\mathbf{a}| = |\mathbf{a}|^2 \tag{3.1.1}$$

which we shall find useful in many following examples.

Notice that if θ is acute then $\mathbf{a}.\mathbf{b}$ is positive, if θ is obtuse then $\mathbf{a}.\mathbf{b}$ is negative, and if θ is a right angle then $\mathbf{a}.\mathbf{b}$ is zero. This last fact is extremely useful, since often we shall want to know whether vectors are *orthogonal* (that is, perpendicular).

Orthogonal Vectors

If vectors **a** and **b** are orthogonal, then from the above we know that $\mathbf{a}.\mathbf{b} = 0$. However, $\mathbf{a}.\mathbf{b} = 0$ does not necessarily imply that the vectors are perpendicular, since one of the vectors could be the zero vector, in which case the angle cannot be defined. So we should say that if $\mathbf{a}.\mathbf{b} = 0$, then $\mathbf{a} = \mathbf{0}$, *or* $\mathbf{b} = \mathbf{0}$, *or* **a** and **b** are orthogonal.

Since $\cos(-\theta) = \cos\theta$,

$$\mathbf{b}.\mathbf{a} = |\mathbf{b}||\mathbf{a}|\cos(-\theta) = |\mathbf{a}||\mathbf{b}|\cos(\theta) = \mathbf{a}.\mathbf{b}$$

and this shows that the scalar product operation is *commutative*.

The scalar product *behaves* like a product in the sense that it is *distributive* over addition. That is,

$$\mathbf{a}.(\mathbf{b}+\mathbf{c}) = \mathbf{a}.\mathbf{b}+\mathbf{a}.\mathbf{c}$$

$$(\mathbf{b}+\mathbf{c}).\mathbf{a} = \mathbf{b}.\mathbf{a}+\mathbf{c}.\mathbf{a}.$$

This is particularly useful when we express vectors in terms of the standard unit vectors **i**, **j** and **k**, for suppose

$$\mathbf{a} = a_1\mathbf{i} + a_2\mathbf{j} + a_3\mathbf{k} \quad \text{and} \quad \mathbf{b} = b_1\mathbf{i} + b_2\mathbf{j} + b_3\mathbf{k}.$$

So,

$$\mathbf{a}.\mathbf{b} = a_1\mathbf{i}.b_1\mathbf{i} + a_2\mathbf{j}.b_1\mathbf{i} + a_3\mathbf{k}.b_1\mathbf{i} + a_1\mathbf{i}.b_2\mathbf{j} + a_2\mathbf{j}.b_2\mathbf{j} + a_3\mathbf{k}.b_2\mathbf{j}$$
$$+ a_1\mathbf{i}.b_3\mathbf{k} + a_2\mathbf{j}.b_3\mathbf{k} + a_3\mathbf{k}.b_3\mathbf{k}.$$

Now $\alpha\mathbf{i}.\beta\mathbf{j} = 0$ whatever the values of α and β, since the vectors are orthogonal. Also $\alpha\mathbf{i}.\beta\mathbf{i} = |\alpha\mathbf{i}||\beta\mathbf{i}|\cos 0 = \alpha\beta$ etc., so that

$$\mathbf{a}.\mathbf{b} = a_1b_1 + a_2b_2 + a_3b_3.$$

Note that

$$\mathbf{a}.\mathbf{a} = a_1^2 + a_2^2 + a_3^2$$

which confirms the result given in equation (3.1.1).

EXERCISE I

Find $\mathbf{a}.\mathbf{b}$ in each of the following cases:

(i) $\mathbf{a} = \mathbf{i} + 2\mathbf{j} + 3\mathbf{k}, \quad \mathbf{b} = -2\mathbf{i} + \mathbf{j} + \mathbf{k},$

(ii) $\mathbf{a} = 5\mathbf{i} - 2\mathbf{j} + 3\mathbf{k}, \quad \mathbf{b} = -2\mathbf{i} + 3\mathbf{j} + 6\mathbf{k}.$

EXERCISE 2

From the following vectors, find as many pairs of orthogonal vectors as you can:

$$\mathbf{a} = \mathbf{i} + 2\mathbf{j} + 3\mathbf{k}, \quad \mathbf{b} = 2\mathbf{i} - \mathbf{j} + \mathbf{k}, \quad \mathbf{c} = 2\mathbf{i} - \mathbf{j}, \quad \mathbf{d} = \mathbf{i} + \mathbf{j} - \mathbf{k}, \quad \mathbf{e} = \mathbf{i} + 2\mathbf{j}.$$

● Example I

Given that $OACB$ is a parallelogram, show that it is a rhombus if and only if its diagonals are at right angles.

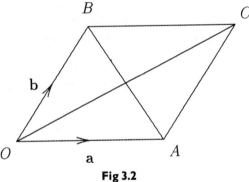

Fig 3.2

SOLUTION

Since we are told that $OACB$ is a parallelogram we know that the opposite sides are both parallel and equal in length, so that if **OA** = **a** and **OB** = **b**, then **BC** = **a** and **AC** = **b**. The diagonals of the parallelogram are OC and BA. From the triangles in the diagram, we can see that

BA = **a** − **b**   and   **OC** = **a** + **b**.

Then

BA.OC = (**a** − **b**).(**a** + **b**)

= **a.a** + **a.b** − **b.a** − **b.b**

= $|\mathbf{a}|^2 - |\mathbf{b}|^2$.

So **BA.OC** = 0 if and only if $|\mathbf{a}| = |\mathbf{b}|$, which means that the diagonals are perpendicular if and only if all the sides are equal in length; that is, if and only if the parallelogram is a rhombus.

3.2 Projections and components

When we see a film, or slides being shown, the pictures are projected onto a screen. In general the clearest images are obtained when the rays meeting the screen are as close to being perpendicular to the screen as possible. When showing slides it is usual to have the line joining the centre of the picture on the screen to the lens of the projector at right angles to the screen. In wide-angle films, to prevent the

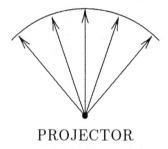

PROJECTOR

Fig 3.3

distortion a flat screen would produce, curved screens are used, and the projector is placed on the axis of the cylinder in which the screen lies, so that all the horizontal rays from the projector meet the screen orthogonally. In this way a better definition of picture is achieved.

In this section we shall be concerned only with orthogonal projections onto either a line or a plane. Consider first the two-dimensional case.

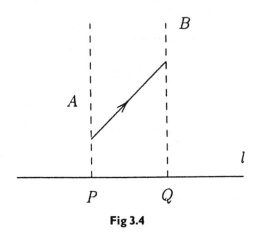

Fig 3.4

Figure 3.4 illustrates the orthogonal projection PQ of the line segment AB onto the line l. The orthogonal projection is the set of all points X on l where the perpendicular to l at X meets the line segment AB. Another way of looking at it is by considering the projection of AB onto l to be the *shadow* of AB on l produced by shining a parallel beam of light orthogonal to l from a position with AB between the source of light and the line l. If instead of the straight line AB we have a curve between A and B we should get the same projection provided the curve did not extend beyond the dashed lines PA and QB perpendicular to l.

The three-dimensional case is similar. Now instead of a line l we can have a plane π, and the projection of the curve AB is the set of all points X in π for which the line orthogonal to π at X meets the curve AB. Again, if we think of the plane as being horizontal, the projection of AB onto the plane is the shadow of AB on the plane in a vertical parallel beam of light.

It is possible to project a curve onto a *line* in $\mathbb{R}^3$. In this case the curve would lie entirely in a plane containing l and at right angles to π. We shall meet this idea again later in the book.

So what has this to do with scalar products? Let us consider the straight line case. We shall consider this in two dimensions first. From Fig 3.5 we see that AB will produce the same projection on l as PB', or in fact any other line segment parallel to AB and having the same length as AB.

Now from triangle $PB'Q$ we get $PQ = PB' \cos \theta = AB \cos \theta$. But if $\hat{\mathbf{c}}$ is a unit vector in the direction of the line l (as shown in the diagram), then

$$\hat{\mathbf{c}} . \mathbf{AB} = 1 . AB . \cos \theta$$

which is the projection of AB onto l. Normally we are only interested in the length of the projection, so if the scalar product is negative we drop the minus sign.

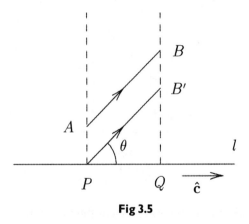

Fig 3.5

Because in practical applications we need to know how much effect a particular vector has in a particular direction, the idea of projections is important.

● *Definition 2*

Given the vectors **a** and **b**, we define the *component of* **a** *in the direction of* **b** to be **a·b̂**.

● *Example 2*

Find the component of **a** in the direction of **b** when

$$\mathbf{a} = 3\mathbf{i} - 2\mathbf{j} + \mathbf{k}, \quad \mathbf{b} = 2\mathbf{i} + 2\mathbf{j} + \mathbf{k}.$$

SOLUTION
$\hat{\mathbf{b}} = \frac{2}{3}\mathbf{i} + \frac{2}{3}\mathbf{j} + \frac{1}{3}\mathbf{k}$, so that $\mathbf{a}\cdot\hat{\mathbf{b}} = 1$. Hence the component of **a** in the direction of **b** is 1.

EXERCISE 3

With the vectors given in the above example, find the component of **b** in the direction of **a**.

Compare this with the idea of components in the directions of **i**, **j** and **k** in Chapter 1. If $\mathbf{a} = a_1\mathbf{i} + a_2\mathbf{j} + a_3\mathbf{k}$, since **i** is already a unit vector, the component of **a** in the direction of **i** is

$$\mathbf{a}\cdot\mathbf{i} = a_1$$

which is as we defined it in Chapter 1.

We can now consider what the scalar product means geometrically. We call the scalar product a *product* since it contains the product of the lengths of the two vectors, but it also contains cos θ.

In Fig 3.6 we see that the projection of **a** onto the direction of **b** is $|\mathbf{a}|\cos\theta$. We could think of **b** as the projection of **b** onto itself, since $|\mathbf{b}|\hat{\mathbf{b}} = \mathbf{b}$. So **a·b** is the product of the components of the two vectors in the direction of **b**. But by similar

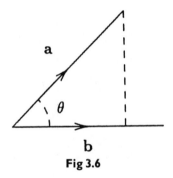

Fig 3.6

reasoning it is the product of the components of the two vectors in the direction of **a**. We could combine these two results into the single statement:

a.b *is the product of the components of* **a** *and* **b** *in the direction of either.*

3.3 Angles from scalar products

Because we have two definitions of **a.b** for those vectors whose components we know, we can use the two expressions to find the angle between the vectors.

Suppose $\mathbf{a} = a_1\mathbf{i} + a_2\mathbf{j} + a_3\mathbf{k}$ and $\mathbf{b} = b_1\mathbf{i} + b_2\mathbf{j} + b_3\mathbf{k}$; then

$$\mathbf{a.b} = |\mathbf{a}||\mathbf{b}|\cos\theta \tag{3.3.1}$$

where θ is the angle between the directions of **a** and **b**, and

$$\mathbf{a.b} = a_1b_1 + a_2b_2 + a_3b_3. \tag{3.3.2}$$

Equating (3.3.1) and (3.3.2) we have

$$|\mathbf{a}||\mathbf{b}|\cos\theta = a_1b_1 + a_2b_2 + a_3b_3$$

and hence

$$\cos\theta = \frac{a_1b_1 + a_2b_2 + a_3b_3}{\sqrt{a_1^2 + a_2^2 + a_3^2}\sqrt{b_1^2 + b_2^2 + b_3^2}}.$$

● *Example 3*

Suppose $\mathbf{a} = 2\mathbf{i} + 2\mathbf{j} - \mathbf{k}$ and $\mathbf{b} = 2\mathbf{i} - 3\mathbf{j} + \mathbf{k}$; then the angle between **a** and **b** is θ, where

$$\cos\theta = \frac{a_1b_1 + a_2b_2 + a_3b_3}{\sqrt{a_1^2 + a_2^2 + a_3^2}\sqrt{b_1^2 + b_2^2 + b_3^2}} = \frac{2.2 + 2.(-3) + (-1).1}{\sqrt{9}\sqrt{14}}.$$

So

$$\cos\theta = \frac{-1}{\sqrt{14}} \quad \Rightarrow \quad \theta \approx 105.5°.$$

EXERCISE 4

Find the angle between the directions of **a** and **b** in the following cases (giving your answer to the nearest degree):

(i) $\mathbf{a} = 3\mathbf{i} - 2\mathbf{j} + \mathbf{k},\quad \mathbf{b} = 2\mathbf{i} + 2\mathbf{j} + \mathbf{k},$

(ii) $\mathbf{a} = 3\mathbf{i} - 2\mathbf{j} + 6\mathbf{k},\quad \mathbf{b} = 2\mathbf{i} + \mathbf{j} - 2\mathbf{k}.$

Direction Cosines

In Chapter 1, we defined the direction cosines of **a** as the components of **â**. We can now see the reasoning behind this terminology since if $\theta_1, \theta_2, \theta_3$ are the angles between the direction of **a** and those of **i, j, k** respectively, then

$$\cos\theta_1 = \frac{a_1.1 + a_2.0 + a_3.0}{1.\sqrt{a_1^2 + a_2^2 + a_3^2}} = \frac{a_1}{|\mathbf{a}|}$$

which is the component of **â** in the direction of **i**. We get similar results for the directions of **j** and **k**. So

$$\cos\theta_1 = \frac{a_1}{|\mathbf{a}|}, \quad \cos\theta_2 = \frac{a_2}{|\mathbf{a}|}, \quad \cos\theta_3 = \frac{a_3}{|\mathbf{a}|}$$

and hence

$$\mathbf{\hat{a}} = (\cos\theta_1, \cos\theta_2, \cos\theta_3).$$

Thus the components of **â** in the directions of **i, j** and **k** are the *cosines of the angles* made by the direction of the vector **a** with these three respective directions.

The Cosine Rule

In Chapter 1, we saw from the triangle of vectors rule that the sides of triangle *ABC* obey the rule $\mathbf{BC} = \mathbf{BA} + \mathbf{AC}$.

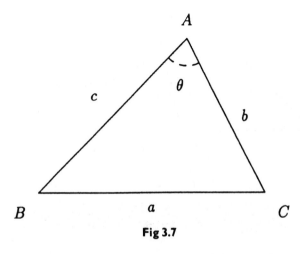

Fig 3.7

Taking the scalar product of each side of this equation with **BC** we get

$$\mathbf{BC}.\mathbf{BC} = (\mathbf{BA} + \mathbf{AC}).\mathbf{BC}$$
$$= (\mathbf{BA} + \mathbf{AC}).(\mathbf{BA} + \mathbf{AC})$$
$$= \mathbf{BA}.\mathbf{BA} + 2\mathbf{BA}.\mathbf{AC} + \mathbf{AC}.\mathbf{AC}.$$

Recall that $\mathbf{AB} = -\mathbf{BA}$ so that we can write the above equation as

$$BC^2 = BA^2 + AC^2 - 2AB.AC\cos\theta,$$

where θ is the angle between the directions of the vectors **AB** and **AC**. With the usual convention that in triangle ABC the sides opposite angle A, B, C are called respectively a, b, c, this reduces to the formula

$$a^2 = c^2 + b^2 - 2cb\cos A$$

which is the standard formula for the *cosine rule*.

EXERCISE 5

Using vector methods, find the lengths of the sides and the angles of triangle ABC, where the coordinates of the points A, B, C are respectively $(1, 2, 3)$, $(4, -3, 2)$, $(2, -1, -3)$.

3.4 Vector equation of a plane

Consider a plane π and a vector **n** which is orthogonal to π. Suppose A is a fixed point on the plane, and let **a** be the position vector of A. Let R, with position vector **r**, be any point on the plane. Now, from Chapter 1, $\mathbf{AR} = \mathbf{r} - \mathbf{a}$, and since **AR** is a vector parallel to the plane π and **n** is orthogonal to π, we have $\mathbf{AR}.\mathbf{n} = (\mathbf{r} - \mathbf{a}).\mathbf{n} = 0$. So the vector equation of the plane passing through the point whose position vector is **a**, perpendicular to the vector **n**, is

$$(\mathbf{r} - \mathbf{a}).\mathbf{n} = 0. \tag{3.4.1}$$

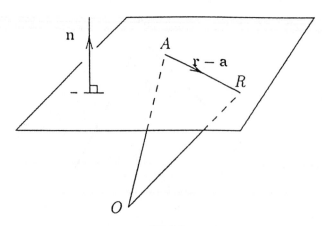

Fig 3.8

This leads to finding the cartesian equation of a plane, for we think of a general point R in $\mathbb{R}^3$ as having coordinates (x, y, z) and position vector $\mathbf{r}$, a given point A with coordinates (a, b, c) and position vector $\mathbf{a}$, and a vector $\mathbf{n} = p\mathbf{i} + q\mathbf{j} + r\mathbf{k}$ normal to the plane. We can write the cartesian equation of the plane as

$$ax + by + cz - (ap + bq + cr) = 0$$

or, writing $d = ap + bq + cr$ we can express the equation of π as

$$ax + by + cz = d. \tag{3.4.2}$$

● Example 4

Suppose a vector normal to the plane π is $\mathbf{i} + 2\mathbf{j} + 2\mathbf{k}$, and suppose π contains the point A whose coordinates are $(2, 1, 5)$. Find the vector equation and the cartesian equation of π.

SOLUTION

The position vector of A is $\mathbf{a} = 2\mathbf{i} + \mathbf{j} + 5\mathbf{k}$ so the vector equation of the plane is

$$\left(\mathbf{r} - \begin{pmatrix} 2 \\ 1 \\ 5 \end{pmatrix} \right) \cdot \begin{pmatrix} 1 \\ 2 \\ 2 \end{pmatrix} = 0$$

and the cartesian equation is

$$x + 2y + 2z - (2 \times 1 + 1 \times 2 + 5 \times 2) = 0$$

or

$$x + 2y + 2z = 14.$$

EXERCISE 6

Find both the vector equation and the cartesian equation of the plane orthogonal to the vector $\mathbf{i} + 3\mathbf{j} - 2\mathbf{k}$ and passing through the point whose coordinates are $(4, -1, 2)$.

3.5 The intersection of two planes

As we have seen above, the cartesian equation of a plane is

$$ax + by + cz = d$$

where at least one of a, b, c is non-zero. Consider the two planes π and π' whose respective equations are

$$ax + by + cz = d \tag{3.5.1}$$

$$a'x + b'y + c'z = d'. \tag{3.5.2}$$

We lose no generality in supposing $a \neq 0$, since at least one of a, b, c is non-zero. Then substituting for x from (3.5.1) into (3.5.2) we get, after a bit of tidying up,

$$\left(b' - \frac{ba'}{a} \right) y + \left(c' - \frac{ca'}{a} \right) z = d' - \frac{da'}{a}. \tag{3.5.3}$$

Suppose the coefficients of y and z are both zero. Then (3.5.3) only has a solution if $d' - da'/a = 0$, since otherwise we should have an equation which had zero on one side and non-zero on the other. This means that there can be no intersection of the planes π and π' if $b' = ba'/a$ and $c' = ca'/a$ unless $d' = da'/a$ also, and if all three of these equations are true then equation (3.5.2) is produced by multiplying both sides of (3.5.1) by a'/a, which means that the two equations (3.5.1) and (3.5.2) represent the same plane. So in this case π and π' are coincident. Now $b' = ba'/a$ and $c' = ca'/a$ imply that $a' : b' : c' = 1 : b/a : c/a = a : b : c$. This means that the planes π and π' are parallel but not coincident if and only if $a' : b' : c' = a : b : c$, but $a' : b' : c' : d' \neq a : b : c : d$.

Going back to (3.5.3), suppose (without loss of generality) that the coefficient of y is non-zero; then we have $y = Az + B$ and $x = Cz + D$ where A, B, C, D are constants, and these two equations define a straight line.

To summarise what we have found here: the two planes π and π' intersect in a straight line unless

(i) they are parallel, in which case $a' : b' : c' = a : b : c$, but $a' : b' : c' : d' \neq a : b : c : d$, or

(ii) they are coincident, in which case $a' : b' : c' : d' = a : b : c : d$.

Example 5

Find the equation of the line of intersection of the planes $x - y - 2z = 3$ and $2x + 3y + 4z = -2$.

SOLUTION
From the first equation we can write $x = y + 2z + 3$, and substituting into the second equation gives

$$2(y + 2z + 3) + 3y + 4z = -2 \quad \text{or} \quad 5y + 8z = -8.$$

This means that $y = -\frac{8}{5}(z + 1)$, so substituting in the expression for x gives

$$x = -\frac{8}{5}(z + 1) + 2z + 3 \quad \text{or} \quad x = \frac{2z + 7}{5}$$

and we can collect together this information to give

$$\frac{x - 7/5}{2} = \frac{y + 8/5}{-8} = \frac{z}{5}$$

which is the set of equations of the line passing through the point $(7/5, -8/5, 0)$ having direction $(2, -8, 5)$. In vector notation this would be expressed as

$$\mathbf{r} = \begin{pmatrix} 7/5 \\ -8/5 \\ 0 \end{pmatrix} + \lambda \begin{pmatrix} 2 \\ -8 \\ 5 \end{pmatrix}.$$

There are a couple of useful checks which can be used here. Firstly we can test whether $(7/5, -8/5, 0)$ really does lie on both planes by substituting into both equations. Now

$$\frac{7}{5} - \frac{(-8)}{5} - 2 \times 0 = 3 \quad \text{and} \quad 2 \times \frac{7}{5} + 3 \times \frac{(-8)}{5} + 4 \times 0 = -2,$$

so our first check is positive. Secondly, if the vectors **n** and **m** are perpendicular to the two planes, then, since l lies in both planes, its direction is perpendicular to both **n** and **m**. In this case

$$\mathbf{n} = \begin{pmatrix} 1 \\ -1 \\ -2 \end{pmatrix} \quad \text{and} \quad \mathbf{m} = \begin{pmatrix} 2 \\ 3 \\ 4 \end{pmatrix}.$$

The vector $\begin{pmatrix} 2 \\ -8 \\ 5 \end{pmatrix}$ is parallel to the line l, and

$$\begin{pmatrix} 1 \\ -1 \\ -2 \end{pmatrix} \cdot \begin{pmatrix} 2 \\ -8 \\ 5 \end{pmatrix} = 0 \quad \text{and} \quad \begin{pmatrix} 2 \\ 3 \\ 4 \end{pmatrix} \cdot \begin{pmatrix} 2 \\ -8 \\ 5 \end{pmatrix} = 0.$$

This verifies that we have the correct direction for the line of intersection and our second check is complete.

EXERCISE 7

Find the equation of the line of intersection (where possible) of the two planes π and π' whose equations are

(i) $x + 2y + z = 2, \quad x - 4y + 3z = 3,$
(ii) $x + 2y + z = 2, \quad 2x + 4y + 2z = 3,$
(iii) $2x - 3y + 2z = 5, \quad 4x - y - z = 3,$
(iv) $x + 2y - 3z = 2, \quad 3x + 6y - 9z = 6.$

Where a line of intersection is found, use the above checks to verify your result.

3.6 The intersection of three planes

We shall consider these systematically. Let us call the three planes π_1, π_2 and π_3, and let $\mathbf{n}_1, \mathbf{n}_2, \mathbf{n}_3$ be vectors normal to these three planes respectively.

Case 1

The three planes are coincident. Then every point on each plane lies on all three planes, so the intersection of the three planes is a plane, namely π_1 $(= \pi_2 = \pi_3)$.

Case 2

If π_1, π_2 and π_3 all contain the same line, but are not all three identical, then the intersection of the three planes is that line. This can happen either when two of the planes coincide and the third is not parallel to these two, or when the normals of the three planes are all distinct but all parallel to a given plane. In this last case the planes fan out from the line of intersection like pages from a very narrow-spined book (see Fig 3.9(ii)).

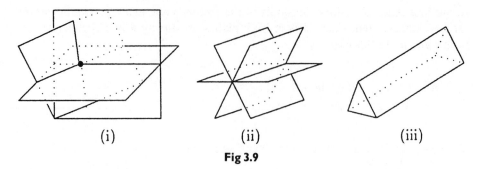

(i)   (ii)   (iii)

Fig 3.9

Case 3

The most general case is when the three planes meet in a single point, and in this case there is no plane which is parallel to all three vectors $\mathbf{n}_1, \mathbf{n}_2, \mathbf{n}_3$ (see Fig 3.9(i)).

Case 4

If any pair of planes is parallel and not coincident, or if the pairs of planes meet in three distinct parallel lines as the rectangular faces of a triangular prism do (see Fig 3.9(iii)), then there is no point lying on all three planes.

The process by which we calculate the intersection of the planes is by solving (or attempting to solve) the three equations simultaneously. We do this by eliminating one of the three variables x, y, z by taking the equations two at a time, thus getting an equation in two variables in each case. Experience should tell us how to solve two linear equations in two unknowns.

◉ Example 6

(i) Consider the three planes whose equations are

$$2x + y + 2z = 2$$
$$2x - y - \ z = 1$$
$$x + y + \ z = 2.$$

By firstly adding the first two equations, and secondly adding the last two equations, and dividing by 3 we get

$$4x + z = 3$$
$$x = 1.$$

Solving these simultaneous equations gives $x = 1$ and $z = -1$, and by substituting into any of the three plane equations we get $y = 2$. So the three planes intersect in the single point $(1, 2, -1)$.

(ii) Consider the three planes whose equations are

$$x + y + 2z = 3$$
$$2x + 2y + 4z = 4$$
$$x - y + 3z = 2.$$

We can see by inspection that the first two planes are parallel (and not coincident) so we immediately know that there is no point which is on all three planes. Thus the intersection of the three planes is the empty set which we denote by $\emptyset$.

Summary

1. The scalar product of the vectors $\mathbf{a}$ and $\mathbf{b}$ is defined as

$$\mathbf{a.b} = |\mathbf{a}||\mathbf{b}| \cos \theta$$

where θ is the angle between the directions of $\mathbf{a}$ and $\mathbf{b}$, or, if

$$\mathbf{a} = a_1\mathbf{i} + a_2\mathbf{j} + a_3\mathbf{k} \quad \text{and} \quad \mathbf{b} = b_1\mathbf{i} + b_2\mathbf{j} + b_3\mathbf{k},$$

$$\mathbf{a.b} = a_1b_1 + a_2b_2 + a_3b_3.$$

2. If $\mathbf{a.b} = 0$ then $\mathbf{a} = \mathbf{0}$ or $\mathbf{b} = \mathbf{0}$ or $\mathbf{a}$ and $\mathbf{b}$ are orthogonal.
3. The length of a vector is connected with the scalar product as follows:

$$|\mathbf{a}|^2 = \mathbf{a.a}.$$

4. The cosine of the angle θ between the directions of $\mathbf{a}$ and $\mathbf{b}$ as defined in 1 is given by

$$\cos \theta = \frac{a_1b_1 + a_2b_2 + a_3b_3}{\sqrt{a_1^2 + a_2^2 + a_3^2}\sqrt{b_1^2 + b_2^2 + b_3^2}}.$$

5. The component of $\mathbf{a}$ in the direction of $\mathbf{b}$ is $\mathbf{a}.\hat{\mathbf{b}}$.
6. $\mathbf{a.b}$ is the product of the components of the two vectors in the direction of either.
7. The equation of a plane is $(\mathbf{r} - \mathbf{a}).\mathbf{n} = 0$ where $\mathbf{n}$ is a vector orthogonal to the plane, and $\mathbf{a}$ is the position vector of a point on the plane.
   If $\mathbf{n} = n_1\mathbf{i} + n_2\mathbf{j} + n_3\mathbf{k}$ and $\mathbf{a} = a_1\mathbf{i} + a_2\mathbf{j} + a_3\mathbf{k}$, the cartesian equation of the plane is

$$n_1x + n_2y + n_3z = n_1a_1 + n_2a_2 + n_3a_3.$$

8. Two planes intersect in a straight line unless (i) they are parallel when they do not intersect at all, or (ii) they are coincident in which case the intersection is the whole plane.
9. The intersection of three planes can be a plane, or a line, or a single point, or the empty set (where there is no point which lies on all three planes).

FURTHER EXERCISES

8. Let $\mathbf{a} = 2\mathbf{i} + \mathbf{j} + 2\mathbf{k}$, $\mathbf{b} = \mathbf{i} + \mathbf{j} + \mathbf{k}$ and $\mathbf{c} = 4\mathbf{i} - 3\mathbf{j} + \mathbf{k}$.
   (i) Find $\mathbf{a.b}$, $\mathbf{b.c}$ and $\mathbf{c.a}$.

(ii) Find the angles between the directions of **a** and **b**, **b** and **c**, **c** and **a** to the nearest degree.

9. Let P, Q, R be the vertices of a triangle where (with the usual convention)

$$\mathbf{p} = 8\mathbf{i} + 3\mathbf{j} + 4\mathbf{k}, \quad \mathbf{q} = 2\mathbf{i} + \mathbf{j} + \mathbf{k}, \quad \mathbf{r} = 5\mathbf{i} - \mathbf{j} + 7\mathbf{k}.$$

Show that $\triangle PQR$ is isosceles and find its angles.

10. Let $\mathbf{a} = 4\mathbf{i} - 2\mathbf{j} + 3\mathbf{k}$ and $\mathbf{b} = 5\mathbf{i} + \mathbf{j} - 2\mathbf{k}$.
    (i) Find the component of **a** in the direction of **b**.
    (ii) Find the component of **b** in the direction of **a**.
    (iii) Multiply your answer to (i) by $|\mathbf{b}|$.
    (iv) Multiply your answer to (ii) by $|\mathbf{a}|$.
    (v) Are you surprised? If so, why? If not, why not?

11. Using the vectors **a** and **b** from Exercise 10, find the direction cosines of $\mathbf{a} + \mathbf{b}$, and of $\mathbf{a} - \mathbf{b}$.

12. (i) Find the equation of the plane which passes through the point $(2,1,5)$ and which is perpendicular to the vector $3\mathbf{i} - \mathbf{j} + 3\mathbf{k}$.
    (ii) Find the equation of the plane which passes through the three points $(1, -2, 5)$, $(2, -4, 1)$ and $(3, 3, -2)$.

13. Find equations of the line of intersection of the planes

$$2x + 4y - 3z = 3 \quad \text{and} \quad 3x - y + 2z = 2.$$

14. Find, where possible, the intersection of the three planes whose equations are given. If the intersection is empty, say why.

(i) $3x - 2y + 2z = 5$
     $4x + y - z = 3$
     $x + y + z = 6$

(ii) $x + 2y - z = 3$
      $2x + 4y - 2z = 5$
      $x - y + z = 1$

(iii) $x + 2y - z = 2$
       $x - y + z = 1$
       $3x + 3y - z = 5$

(iv) $x + y - z = 1$
      $3x + y + 2z = 6$
      $2x + 3z = 4.$

15. Describe geometrically the following sets of points of $\mathbb{R}^3$, where **r** is the position vector of a variable point R, and **a** is the position vector of a fixed point A:
    (i) $\mathbf{r}.\mathbf{a} = 0$,
    (ii) $(\mathbf{r} - \mathbf{a}).\mathbf{a} = 0$,
    (iii) $(\mathbf{r} - \mathbf{a}).\mathbf{r} = 0$,
    (iv) $\mathbf{r}.\mathbf{r} = \mathbf{a}.\mathbf{a}$,
    (v) $(\mathbf{r} - \mathbf{a}).(\mathbf{r} - \mathbf{a}) = \mathbf{a}.\mathbf{a}$.

4 • Vector Products

4..1 Definition and geometrical description

We have defined a *scalar product* of vectors, where two vectors are combined to produce a scalar. We now consider a product which combines two vectors to give a vector.

• Definition 1

The *vector product* of the two vectors **a** and **b** is defined to be **a** × **b** where

$$\mathbf{a} \times \mathbf{b} = |\mathbf{a}||\mathbf{b}| \sin\theta \hat{\mathbf{n}}$$

where θ is the angle between the directions of **a** and **b** and $\hat{\mathbf{n}}$ is a unit vector orthogonal to both **a** and **b** whose sense is given by a right hand screw rule on the triad **a**, **b**, $\hat{\mathbf{n}}$, as shown in Fig 4.1. (We discussed right hand triads in Chapter 1, but there we were concerned with the special case where the three directions were mutually perpendicular. However, the principle is the same.)

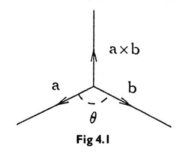

Fig 4.1

Geometrically $|\mathbf{a} \times \mathbf{b}|$ is the area of the parallelogram whose sides represent the two vectors **a** and **b** in both magnitude and direction as shown in Fig 4.2.

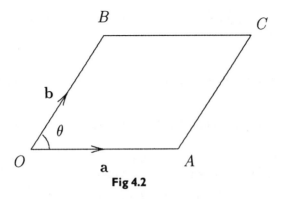

Fig 4.2

Because the sine of a zero angle is zero,

$$\mathbf{i} \times \mathbf{i} = \mathbf{j} \times \mathbf{j} = \mathbf{k} \times \mathbf{k} = \mathbf{0}.$$

It is very important to notice that this gives the zero *vector* and not the number 0, since the vector product of two vectors is always a vector.

From the definition we find that, unlike the scalar product, the vector product is *not* commutative. In fact we have

$$\mathbf{a} \times \mathbf{b} = -\mathbf{b} \times \mathbf{a}$$

as, if the triads $\mathbf{a}$, $\mathbf{b}$, $\hat{\mathbf{n}}_1$ and $\mathbf{b}$, $\mathbf{a}$, $\hat{\mathbf{n}}_2$ are both right hand triads, then

$$\hat{\mathbf{n}}_1 = -\hat{\mathbf{n}}_2.$$

This means that

$$\mathbf{i} \times \mathbf{j} = \mathbf{k} = -\mathbf{j} \times \mathbf{i}, \quad \mathbf{j} \times \mathbf{k} = \mathbf{i} = -\mathbf{k} \times \mathbf{j}, \quad \mathbf{k} \times \mathbf{i} = \mathbf{j} = -\mathbf{i} \times \mathbf{k}.$$

This is a good example of positive and negative cyclic expressions.

The Baked Bean Tin Method

Imagine $1, 2, 3$ being equally spaced on a strip of paper stuck round a baked bean tin as shown in Fig 4.3, and take the direction the tin is turned to move it from the position showing 1 at the front to the position showing 2 at the front as the *positive* direction. The opposite direction will be considered as *negative*.

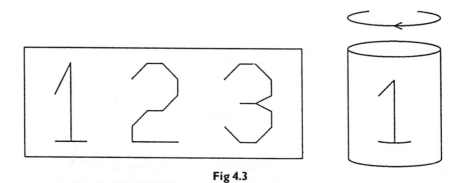

Fig 4.3

Then a positive sign will be attached to any block of three taken from the sequence

1 2 3 1 2 3 1 ...

and a negative sign will be attached to any block of three taken from the sequence

1 3 2 1 3 2 1 ....

This method is also useful in calculating the determinant of a matrix – something we shall need to do very soon.

As with the scalar product, we have a special result when we express our vectors $\mathbf{a}$ and $\mathbf{b}$ in terms of the vectors $\mathbf{i}$, $\mathbf{j}$ and $\mathbf{k}$. Suppose

$$\mathbf{a} = a_1\mathbf{i} + a_2\mathbf{j} + a_3\mathbf{k} \quad \text{and} \quad \mathbf{b} = b_1\mathbf{i} + b_2\mathbf{j} + b_3\mathbf{k}.$$

Then,

$$\mathbf{a} \times \mathbf{b} = a_1\mathbf{i} \times b_1\mathbf{i} + a_1\mathbf{i} \times b_2\mathbf{j} + a_1\mathbf{i} \times b_3\mathbf{k} + a_2\mathbf{j} \times b_1\mathbf{i} + a_2\mathbf{j} \times b_2\mathbf{j}$$
$$+ a_2\mathbf{j} \times b_3\mathbf{k} + a_3\mathbf{k} \times b_1\mathbf{i} + a_3\mathbf{k} \times b_2\mathbf{j} + a_3\mathbf{k} \times b_3\mathbf{k}.$$

From the results above this reduces to

$$\mathbf{a} \times \mathbf{b} = (a_2b_3 - a_3b_2)\mathbf{i} + (a_3b_1 - a_1b_3)\mathbf{j} + (a_1b_2 - a_2b_1)\mathbf{k}.$$

We can define the *determinant* of a matrix

$$\begin{pmatrix} p_1 & p_2 & p_3 \\ q_1 & q_2 & q_3 \\ r_1 & r_2 & r_3 \end{pmatrix}$$

as

$$\begin{vmatrix} p_1 & p_2 & p_3 \\ q_1 & q_2 & q_3 \\ r_1 & r_2 & r_3 \end{vmatrix} = p_1(q_2r_3 - q_3r_2) + p_2(q_3r_1 - q_1r_3) + p_3(q_1r_2 - q_2r_1)$$

$$= p_1q_2r_3 + p_2q_3r_1 + p_3q_1r_2 - p_1q_3r_2 - p_2q_1r_3 - p_3q_2r_1.$$

Note that, if we keep the order of p, q, r fixed, the subscripts 123, 231, 312 take a $+$ sign and the subscripts 132, 213, 321 take a $-$ sign. These agree with the positive and negative blocks from the baked bean tin sequence.

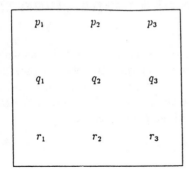 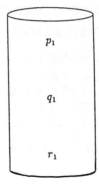

Fig. 4.4

This is where the baked bean tin method comes into its own again. If we write $p_1, p_2, p_3, \ldots, r_3$ on a label as shown in Fig 4.4 and stick it round our baked bean tin, then as we turn the tin in a positive direction we shall see that label repeated as shown; then if we start at the top of the label enclosed in solid lines, the diagonals in the positive direction will give positive products $p_1q_2r_3$, $p_2q_3r_1$, $p_3q_1r_2$ and the diagonals in the negative direction will give the negative products $p_1q_3r_2$, $p_2q_1r_3$, $p_3q_2r_1$ as shown in Fig 4.5.

From above it will be seen that we can write $\mathbf{a} \times \mathbf{b}$ as

$$\mathbf{a} \times \mathbf{b} = \begin{vmatrix} \mathbf{i} & \mathbf{j} & \mathbf{k} \\ a_1 & a_2 & a_3 \\ b_1 & b_2 & b_3 \end{vmatrix}.$$

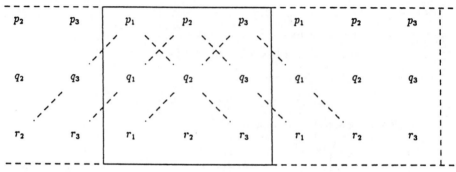

Fig 4.5

Find the vector product $\mathbf{a} \times \mathbf{b}$ in each of the following cases:
  (i) $\mathbf{a} = 2\mathbf{i} + 3\mathbf{j} - \mathbf{k}$, $\mathbf{b} = 3\mathbf{i} - 2\mathbf{j} + 2\mathbf{k}$
  (ii) $\mathbf{a} = \mathbf{i} - 3\mathbf{j} + 5\mathbf{k}$, $\mathbf{b} = \mathbf{i} - 2\mathbf{j} + 4\mathbf{k}$
  (iii) $\mathbf{a} = \mathbf{i} + 3\mathbf{j} - 2\mathbf{k}$, $\mathbf{b} = 3\mathbf{i} - \mathbf{j}$.

We now consider the applications of the vector product.

4.2 Vector equation of a plane given three points on it

The vector product is a useful tool. It is a means of directly calculating a vector perpendicular to any two given vectors. Suppose the plane π passes through the points A, B and C whose respective position vectors are $\mathbf{a}$, $\mathbf{b}$ and $\mathbf{c}$. Then two vectors parallel to the plane are $\mathbf{b} - \mathbf{a}$ and $\mathbf{c} - \mathbf{a}$. Thus a vector normal to the plane is $(\mathbf{b} - \mathbf{a}) \times (\mathbf{c} - \mathbf{a})$, so the equation of the plane becomes

$$(\mathbf{r} - \mathbf{a}).((\mathbf{b} - \mathbf{a}) \times (\mathbf{c} - \mathbf{a})) = 0.$$

Example I

Find the equation of the plane passing through the points A, B and C whose coordinates are $(1, 2, 3)$, $(2, -1, 2)$ and $(1, -1, 1)$.

SOLUTION
With the usual notation

$$\mathbf{b} - \mathbf{a} = \begin{pmatrix} 2 \\ -1 \\ 2 \end{pmatrix} - \begin{pmatrix} 1 \\ 2 \\ 3 \end{pmatrix} = \begin{pmatrix} 1 \\ -3 \\ -1 \end{pmatrix}$$

and

$$\mathbf{c} - \mathbf{a} = \begin{pmatrix} 1 \\ -1 \\ 1 \end{pmatrix} - \begin{pmatrix} 1 \\ 2 \\ 3 \end{pmatrix} = \begin{pmatrix} 0 \\ -3 \\ -2 \end{pmatrix}$$

then

$$(\mathbf{b} - \mathbf{a}) \times (\mathbf{c} - \mathbf{a}) = \begin{pmatrix} 3 \\ 2 \\ -3 \end{pmatrix}$$

so the vector equation of π becomes

$$\left(\mathbf{r} - \begin{pmatrix} 1 \\ 2 \\ 3 \end{pmatrix} \right) \cdot \begin{pmatrix} 3 \\ 2 \\ -3 \end{pmatrix} = 0$$

and the cartesian equation

$$3x + 2y - 3z + 2 = 0.$$

We can easily verify that this is correct by checking that the coordinates of the three points satisfy the equation.

EXERCISE 2

Find the vector equation and the cartesian equation of the plane which passes through the points $(1, -2, 2)$, $(3, -1, 4)$, $(-2, 1, 1)$. Check your results by using the method described in Example 1.

4.3 Distance of a point from a line

If we wish to find the shortest distance from a point to a line, then this distance will be measured along the perpendicular from the point to the line, since if P is the given point, Q is the point on the line l for which PQ is perpendicular to l, and A is any point on l; then PQA is a right-angled triangle with the right angle at Q and hypotenuse AP. Since the hypotenuse is always the longest side of a right-angled triangle, it follows that the distance from P of any point other than Q on the line l is always longer than the distance PQ. Hence the shortest distance of a point from a line is the *perpendicular* distance of the point from the line.

In Fig 4.6, the perpendicular distance from the point P to the line l is PQ. Let A be a point on the line l, and suppose the direction of l is given by the unit vector $\mathbf{u}$. Then the distance of P from l is

$$PQ = AP \sin \theta = |\mathbf{AP} \times \mathbf{u}|.$$

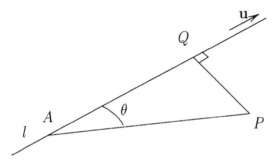

Fig 4.6

Example 2

Find the distance of the point $(4, 2, -1)$ from the line

$$\mathbf{r} = \begin{pmatrix} 2 \\ 3 \\ 2 \end{pmatrix} + \lambda \begin{pmatrix} 1 \\ 2 \\ -2 \end{pmatrix}.$$

SOLUTION

A point A on the line is $(2,3,2)$, so $\mathbf{AP} = \begin{pmatrix} 2 \\ -1 \\ -3 \end{pmatrix}$ and a unit vector in the direction of the line is $\begin{pmatrix} 1/3 \\ 2/3 \\ -2/3 \end{pmatrix}$. Thus the distance of P from the line is

$$\left| \begin{pmatrix} 2 \\ -1 \\ -3 \end{pmatrix} \times \begin{pmatrix} 1/3 \\ 2/3 \\ -2/3 \end{pmatrix} \right| = \left| \frac{1}{3} \begin{pmatrix} 8 \\ 1 \\ 5 \end{pmatrix} \right| = \sqrt{10}.$$

EXERCISE 3

Find the distance of the point $(2, 1, 5)$ from the line

$$\mathbf{r} = \begin{pmatrix} -2 \\ 3 \\ 2 \end{pmatrix} + \lambda \begin{pmatrix} 1 \\ -1 \\ 1 \end{pmatrix}.$$

4.4 Distance between two lines

We found in Chapter 2 that lines could either intersect or lie in parallel planes, and as an example we saw that lines on opposite faces of a cuboid never intersect. However, even if lines do not intersect it is often useful to know how close they get to one another. For instance, if two aircraft are flying in the same region, it is vital to know that there will always be a safe distance between them, so we need to know what their closest distance will be if they continue on their current flight paths.

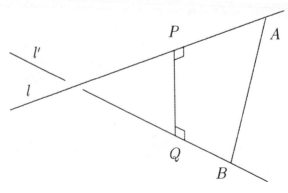

Fig 4.7

Let the two lines be l and l', and let their respective equations be

$$\mathbf{r} = \mathbf{a} + \lambda\mathbf{m} \quad \text{and} \quad \mathbf{r} = \mathbf{b} + \mu\mathbf{n}.$$

Suppose the point P with position vector $\mathbf{p}$ lies on l and the point Q with position vector $\mathbf{q}$ lies on l' in such a way that PQ is the shortest line segment joining a point of l to a point of l'; then PQ is perpendicular to l since it is the shortest distance from Q to l, and PQ is perpendicular to l' since it is the shortest distance of P from l'. This means that PQ is perpendicular to both l and l'.

Now we may not know the position vectors of P and Q, but we do know that since PQ is perpendicular to both l and l', its direction is that of the vector $\mathbf{m} \times \mathbf{n}$, provided this is not the zero vector, and a unit vector in the direction of PQ is

$$\frac{\mathbf{m} \times \mathbf{n}}{|\mathbf{m} \times \mathbf{n}|}.$$

However, from the equations of the lines we know that the point A with position vector $\mathbf{a}$ lies on l, and B with position vector $\mathbf{b}$ lies on l'. Since the lines l and l' lie in parallel planes the distance between them is the component of the vector $\mathbf{a} - \mathbf{b}$ in the direction of the vector $\mathbf{PQ}$, and this is

$$\left| (\mathbf{a} - \mathbf{b}) \cdot \frac{\mathbf{m} \times \mathbf{n}}{|\mathbf{m} \times \mathbf{n}|} \right|.$$

The exceptional case when $\mathbf{m} \times \mathbf{n} = \mathbf{0}$ occurs when the lines l and l' are parallel. In this case we can take any point on the line l and find its distance from l' as shown in Section 4.3, as the distance of any point on one line from the other line is constant.

EXERCISE 4

Find the distance between
(i) the lines whose equations are

$$\mathbf{r} = \begin{pmatrix} 1 \\ -2 \\ 3 \end{pmatrix} + \lambda \begin{pmatrix} 2 \\ 1 \\ 2 \end{pmatrix} \quad \text{and} \quad \mathbf{r} = \begin{pmatrix} 4 \\ -1 \\ 1 \end{pmatrix} + \mu \begin{pmatrix} 1 \\ 3 \\ -1 \end{pmatrix},$$

(ii) the lines whose equations are

$$\mathbf{r} = \begin{pmatrix} 1 \\ 5 \\ 4 \end{pmatrix} + \lambda \begin{pmatrix} 2 \\ 1 \\ 2 \end{pmatrix} \quad \text{and} \quad \mathbf{r} = \begin{pmatrix} -2 \\ 1 \\ 3 \end{pmatrix} + \mu \begin{pmatrix} 4 \\ 2 \\ 4 \end{pmatrix}.$$

4.5 The intersection of two planes

We saw in Chapter 3 that if two planes are neither coincident nor parallel, then they meet in a straight line. We now have a much neater way of finding the equation of this line.

Suppose the planes π and π' are neither coincident nor parallel, and let their line of intersection be l. Suppose also that the vector equations of the two planes are

$$(\mathbf{r} - \mathbf{a}) \cdot \mathbf{n} = 0 \quad \text{and} \quad (\mathbf{r} - \mathbf{b}) \cdot \mathbf{m} = 0.$$

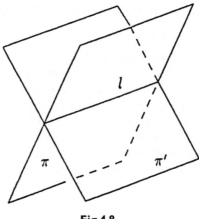

Fig 4.8

Then since **n** is orthogonal to π it is perpendicular to any line in π. In particular it is perpendicular to l. Similarly **m** is perpendicular to l. This means that l is perpendicular to both **n** and **m**, and hence it is parallel to $\mathbf{n} \times \mathbf{m}$. Suppose **c** is a point lying on both π and π'; then the vector equation of l is

$$\mathbf{r} = \mathbf{c} + \lambda(\mathbf{n} \times \mathbf{m}).$$

Example 3

Find the line of intersection of the two planes π and π' whose respective cartesian equations are

$$2x - y + z = 2 \quad \text{and} \quad 3x + y - 3z = 4.$$

SOLUTION

Vectors perpendicular to π and π' respectively are $\mathbf{n} = \begin{pmatrix} 2 \\ -1 \\ 1 \end{pmatrix}$ and $\mathbf{m} = \begin{pmatrix} 3 \\ 1 \\ -3 \end{pmatrix}$, so a vector in the direction of the line of intersection is

$$\mathbf{n} \times \mathbf{m} = \begin{pmatrix} 2 \\ 9 \\ 5 \end{pmatrix}.$$

We need to find some point which lies on both lines. Since neither plane is parallel to $z = 0$, we can choose a point of intersection whose z-coordinate is zero. This will mean that the other coordinates must satisfy both

$$2x - y = 2 \quad \text{and} \quad 3x + y = 4.$$

By adding these we get $5x = 6$, so $x = 6/5$, and substituting back gives $y = 2/5$. Thus a point on the line of intersection is $(6/5, 2/5, 0)$, and it is easy to check that these coordinates do indeed satisfy the equations of both π and π'. Thus the line of intersection of π and π' has vector equation

$$\mathbf{r} = \begin{pmatrix} 6/5 \\ 2/5 \\ 0 \end{pmatrix} + \lambda \begin{pmatrix} 2 \\ 9 \\ 5 \end{pmatrix}.$$

If one of the planes had been parallel to $z = 0$, then its equation would have been of the form $z = k$ and we should take this value of z in the other equation and any values of x and y which satisfy the equation thus formed.

Example 4

Find a point on the line of intersection of the planes whose equations are $z = 2$ and $x + y + z = 3$.

SOLUTION

Substituting $z = 2$ in the second equation gives $x + y = 1$. $x = 1$, $y = 0$ satisfies this, so a point on the line of intersection is $(1, 0, 2)$.

EXERCISE 5

Find the line of intersection of the two planes whose equations are given:
  (i) $x + 2y - 3z = 4$ and $2x - y + 2z = 3$,
  (ii) $x + 2y - 3z = 4$ and $z = 1$.

4.6 Triple scalar product

This product takes three vectors and combines them to produce a scalar:

  $\mathbf{a}.(\mathbf{b} \times \mathbf{c})$.

Let us consider what this means in geometrical terms. In Fig 4.9 we have a parallelepiped (which according to my dictionary should be pronounced as it would be if it were written as two words *parallel epiped*, that is parallelepiped with the stress on the italic *e*), whose faces are all parallelograms, and three edges

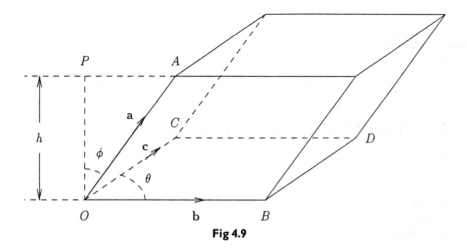

Fig 4.9

OA, OB, OC from one vertex which are represented by the vectors **a**, **b** and **c** as shown. Then $|OA \cos \phi|$ is the *height* of the parallelepiped, and the area of the base is $|OB.OC. \sin \theta|$. Recall that

$$\mathbf{b} \times \mathbf{c} = |\mathbf{b}||\mathbf{c}| \sin \theta \hat{\mathbf{n}}$$

where $\hat{\mathbf{n}}$ is a unit vector in the direction of OP on the diagram, that is perpendicular to the plane $OBDC$. This means that

$$\mathbf{a}.(\mathbf{b} \times \mathbf{c}) = |OA \cos \phi.OB.OC. \sin \theta| \hat{\mathbf{n}},$$

and since $\hat{\mathbf{n}}$ is a unit vector, the magnitude of this vector is the height of the parallelepiped multiplied by the area of its base, that is the *volume* of the parallelepiped.

This can be summarised as follows:

> *The volume of a parallelepiped, three of whose sides are represented in both magnitude and direction by the vectors* **a, b, c**, *is*
>
> $|\mathbf{a}.(\mathbf{b} \times \mathbf{c})|$.

Since the volume of a parallelepiped is independent of which face we take as the *base* it is clear that the triple scalar product with **a, b, c** taken in any order will give us the same magnitude, but may not always give us the same number. To investigate further we look at the triple scalar product in terms of the components of the vectors.

Let $\mathbf{a} = a_1\mathbf{i} + a_2\mathbf{j} + a_3\mathbf{k}$, $\mathbf{b} = b_1\mathbf{i} + b_2\mathbf{j} + b_3\mathbf{k}$, $\mathbf{c} = c_1\mathbf{i} + c_2\mathbf{j} + c_3\mathbf{k}$; then

$$\mathbf{b} \times \mathbf{c} = (b_2c_3 - b_3c_2)\mathbf{i} + (b_3c_1 - b_1c_3)\mathbf{j} + (b_1c_2 - b_2c_1)\mathbf{k}$$

so that

$$\mathbf{a}.(\mathbf{b} \times \mathbf{c}) = a_1b_2c_3 - a_1b_3c_2 + a_2b_3c_1 - a_2b_1c_3 + a_3b_1c_2 - a_3b_2c_1$$

which can be written as

$$\begin{vmatrix} a_1 & a_2 & a_3 \\ b_1 & b_2 & b_3 \\ c_1 & c_2 & c_3 \end{vmatrix}.$$

EXERCISE 6

Show that

$$\mathbf{a}.(\mathbf{b} \times \mathbf{c}) = \mathbf{b}.(\mathbf{c} \times \mathbf{a}) = \mathbf{c}.(\mathbf{a} \times \mathbf{b}) = -\mathbf{a}.(\mathbf{c} \times \mathbf{b}), \text{ etc.} \qquad (4.6.1)$$

4.7 Triple vector product

Being a *vector* product, three vectors are combined to produce a vector:

$$\mathbf{a} \times (\mathbf{b} \times \mathbf{c}) = (\mathbf{a}.\mathbf{c})\mathbf{b} - (\mathbf{a}.\mathbf{b})\mathbf{c}. \qquad (4.7.1)$$

Because $\mathbf{b} \times \mathbf{c}$ is a vector **d**, say, which is perpendicular to both **b** and **c**, and $\mathbf{a} \times \mathbf{d}$ is perpendicular to **a** and **d**, then it is a vector in the plane of **b** and **c**, perpendicular to **a**.

EXERCISE 7

Let $\mathbf{a} = \begin{pmatrix} 1 \\ 2 \\ 3 \end{pmatrix}$, $\mathbf{b} = \begin{pmatrix} 2 \\ -1 \\ 2 \end{pmatrix}$ and $\mathbf{c} = \begin{pmatrix} 4 \\ 1 \\ 1 \end{pmatrix}$.

(i) Find $\mathbf{a} \times \mathbf{b}$ and $\mathbf{a} \times \mathbf{c}$ and show that

$$(\mathbf{a} \times \mathbf{b}) \times (\mathbf{a} \times \mathbf{c}) = \lambda\mathbf{a} \qquad (4.7.2)$$

for some real value of λ, and find this value λ.

(ii) Show that equation (4.7.2) is true for any vectors $\mathbf{a}$, $\mathbf{b}$ and $\mathbf{c}$, and find the value of λ in terms of these three vectors.

Summary

1. $\mathbf{a} \times \mathbf{b} = |\mathbf{a}||\mathbf{b}| \sin \theta \, \hat{\mathbf{n}}$.

2. $\mathbf{a} \times \mathbf{b} = \begin{vmatrix} \mathbf{i} & \mathbf{j} & \mathbf{k} \\ a_1 & a_2 & a_3 \\ b_1 & b_2 & b_3 \end{vmatrix}$.

3. The vector equation of the plane passing through the points with position vectors $\mathbf{a}$, $\mathbf{b}$, $\mathbf{c}$ is

    $$(\mathbf{r} - \mathbf{a}).((\mathbf{b} - \mathbf{a}) \times (\mathbf{c} - \mathbf{a})) = 0.$$

4. The distance of a point P from a line passing through A parallel to the unit vector $\mathbf{u}$ is

    $$|\mathbf{AP} \times \mathbf{u}|.$$

5. The distance between the lines $\mathbf{r} = \mathbf{a} + \lambda\mathbf{n}$ and $\mathbf{r} = \mathbf{b} + \mu\mathbf{m}$ is

    $$\left| (\mathbf{a} - \mathbf{b}).\frac{\mathbf{m} \times \mathbf{n}}{|\mathbf{m} \times \mathbf{n}|} \right|.$$

6. The equation of the line of intersection of the planes $(\mathbf{r} - \mathbf{a}).\mathbf{n} = 0$ and $(\mathbf{r} - \mathbf{b}).\mathbf{m} = 0$ is

    $$\mathbf{r} = \mathbf{c} + \lambda(\mathbf{n} \times \mathbf{m})$$

    where $\mathbf{c}$ is a point on the intersection of the two planes.

7. *Triple scalar product*:

    $$\mathbf{a}.(\mathbf{b} \times \mathbf{c}) = a_1 b_2 c_3 - a_1 b_3 c_2 + a_2 b_3 c_1 - a_2 b_1 c_3 + a_3 b_1 c_2 - a_3 b_2 c_1$$

    $$= \begin{vmatrix} a_1 & a_2 & a_3 \\ b_1 & b_2 & b_3 \\ c_1 & c_2 & c_3 \end{vmatrix}.$$

8. *Triple vector product*:

    $$\mathbf{a} \times (\mathbf{b} \times \mathbf{c}) = (\mathbf{a}.\mathbf{c})\mathbf{b} - (\mathbf{a}.\mathbf{b})\mathbf{c}.$$

FURTHER EXERCISES

8. Find $\mathbf{a} \times \mathbf{b}$ in the following cases:
   (i) $\mathbf{a} = 4\mathbf{i} - \mathbf{j} + 3\mathbf{k}, \mathbf{b} = 2\mathbf{i} + 2\mathbf{j} + \mathbf{k}$,
   (ii) $\mathbf{a} = 2\mathbf{i} + 3\mathbf{j} - 6\mathbf{k}, \mathbf{b} = \mathbf{i} - 2\mathbf{j} + 2\mathbf{k}$.

9. Find the volume of the parallelepiped which has one vertex at the origin O, and edges OA, OB, OC where A, B, C have position vectors

$$\mathbf{a} = \mathbf{i} - \mathbf{j} + \mathbf{k}, \quad \mathbf{b} = 2\mathbf{i} - 3\mathbf{j} + 6\mathbf{k}, \quad \mathbf{c} = -3\mathbf{i} + 4\mathbf{j} + 7\mathbf{k}.$$

10. Find the distance of the point P, whose coordinates are $(3, -2, 1)$, from the line l which passes through the points A and B whose respective coordinates are $(1, 0, 1)$ and $(2, -1, 0)$. What are the coordinates of the point Q on the line l which is closest to P?

11. Find the distance between the two lines l and m, where
    (i) l passes through $(1,2,3)$ and $(2, 1, -2)$, and m passes through $(1,1,1)$ and $(-2, 3, 1)$,
    (ii) l passes through the origin and is parallel to the vector $2\mathbf{i} - \mathbf{j} + 2\mathbf{k}$, and m passes through the point $(3,1,1)$ and is parallel to the vector $\mathbf{i} - \mathbf{j} + \mathbf{k}$.

12. Find both the vector equation and the cartesian set of equations of the line of intersection of the planes

$$2x + 3y + z = 6 \quad \text{and} \quad 3x - y + 3z = 5.$$

13. In each of the following cases, find the equations of the intersection of π with π', π' with π'', and π'' with π.
    (i) $\pi : 2x + 3y - 4z = 4$, $\pi' : x + y - z = 2$, $\pi'' : 2x - y + z = 1$
    (ii) $\pi : x + y + z = 4$, $\pi' : 2x - y + 3z = 2$, $\pi'' : 4x + y + 5z = 1$.
    Explain why in (i) the three planes intersect in a single point, and why in (ii) they do not.

14. Let

$$\mathbf{a} = 2\mathbf{i} - \mathbf{j} + 3\mathbf{k}, \quad \mathbf{b} = 2\mathbf{i} - \mathbf{j} + 5\mathbf{k}, \quad \mathbf{c} = 3\mathbf{i} + 2\mathbf{j} - 4\mathbf{k}.$$

   Find
   (i) $\mathbf{a}.(\mathbf{b} \times \mathbf{c})$
   (ii) $\mathbf{a} \times (\mathbf{b} \times \mathbf{c})$
   (iii) $(\mathbf{a} \times \mathbf{c}) \times (\mathbf{b} \times \mathbf{c})$.

15. Simplify the following expressions:
    (i) $(\mathbf{i} \times \mathbf{j}).\{(\mathbf{i} \times \mathbf{k}) \times (\mathbf{j} \times \mathbf{k})\}$,
    (ii) $(\mathbf{i} \times \mathbf{j}) \times \{(\mathbf{i} \times \mathbf{k}) \times (\mathbf{j} \times \mathbf{k})\}$.

16. Find the other vertices of a cube which has three of its vertices at the points $(0, 0, 0), (2, 1, 2), (1, 2, -2)$. (There are two possible answers for this.) Find also the equations of the six planes, each of which contains a face of the cube.

5 • The Vector Spaces $\mathbb{R}^2$ and $\mathbb{R}^3$, Linear Combinations and Bases

5.1 The vector space $\mathbb{R}^n$

Although this book is mainly concerned with vectors in $\mathbb{R}^2$ and $\mathbb{R}^3$ it is useful to consider the case $\mathbb{R}^n$ if only to consider the two cases $\mathbb{R}^2$ and $\mathbb{R}^3$ at the same time. The ideas we have already met simply extend from the 2-vector and 3-vector to the n-vector.

Whereas in $\mathbb{R}^2$ a vector can be written in the form $\begin{pmatrix} a_1 \\ a_2 \end{pmatrix}$ and a vector in $\mathbb{R}^3$ in the form $\begin{pmatrix} a_1 \\ a_2 \\ a_3 \end{pmatrix}$, we can consider a vector in $\mathbb{R}^n$ to be of the form $\begin{pmatrix} a_1 \\ \vdots \\ a_n \end{pmatrix}$ where there are n entries in the column. (In some contexts, $\mathbb{R}^n$ is considered as the set of *points* with coordinates $(a_1, \ldots, a_n)$, where $a_1, \ldots, a_n \in \mathbb{R}$, but in this context we think of $\mathbb{R}^n$ as the set of position vectors of such points. However, the two notions are equivalent.)

Properties of addition and scalar multiplication work for vectors in $\mathbb{R}^n$ in the following way:

$$\begin{pmatrix} a_1 \\ \vdots \\ a_n \end{pmatrix} + \begin{pmatrix} b_1 \\ \vdots \\ b_n \end{pmatrix} = \begin{pmatrix} a_1 + b_1 \\ \vdots \\ a_n + b_n \end{pmatrix} \qquad \text{and} \qquad \alpha \begin{pmatrix} a_1 \\ \vdots \\ a_n \end{pmatrix} = \begin{pmatrix} \alpha a_1 \\ \vdots \\ \alpha a_n \end{pmatrix}.$$

Similarly the length of the vector $\mathbf{a} = \begin{pmatrix} a_1 \\ \vdots \\ a_n \end{pmatrix}$ is $|\mathbf{a}| = \sqrt{a_1^2 + \ldots + a_n^2}$, so a unit vector in the direction of $\mathbf{a}$ is

$$\hat{\mathbf{a}} = \frac{1}{\sqrt{a_1^2 + \ldots + a_n^2}} \begin{pmatrix} a_1 \\ \vdots \\ a_n \end{pmatrix}.$$

The scalar product of two vectors in $\mathbb{R}^n$ is defined as

$$\begin{pmatrix} a_1 \\ \vdots \\ a_n \end{pmatrix} \cdot \begin{pmatrix} b_1 \\ \vdots \\ b_n \end{pmatrix} = a_1 b_1 + \ldots a_n b_n$$

and we say that two non-zero vectors **a** and **b** are orthogonal in $\mathbb{R}^n$ if and only if their scalar product **a.b** is zero.

Note that all these definitions agree with the definitions given earlier for the $\mathbb{R}^2$ and $\mathbb{R}^3$ cases by putting $n = 2$ and $n = 3$ respectively.

The algebra of vectors has many results analogous to ordinary algebra, but there are some differences too, so it is useful to collect together the properties in a tidy list.

For all vectors **u**, **v**, **w** in $\mathbb{R}^n$ and all real numbers α and β:

(a.1) $\mathbf{u} + \mathbf{v} \in \mathbb{R}^n$.
    (This means that $\mathbb{R}^n$ is closed under vector addition.)

(a.2) $\mathbb{R}^n$ contains a zero vector **0** such that $\mathbf{v} + \mathbf{0} = \mathbf{0} + \mathbf{v} = \mathbf{v}$.

(a.3)   For each **v** in $\mathbb{R}^n$ there is a vector $(-\mathbf{v})$ in $\mathbb{R}^n$ such that $(-\mathbf{v}) + \mathbf{v} = \mathbf{v} + (-\mathbf{v}) = \mathbf{0}$.
    (Every vector in $\mathbb{R}^n$ has an additive inverse in $\mathbb{R}^n$.)

(a.4) $\mathbf{v} + (\mathbf{w} + \mathbf{x}) = (\mathbf{v} + \mathbf{w}) + \mathbf{x}$.
    (Vector addition is associative.)

(a.5) $\mathbf{v} + \mathbf{w} = \mathbf{w} + \mathbf{v}$.
    (Vector addition is commutative.)

(m.1) $\alpha\mathbf{v} \in \mathbb{R}^n$.
    ($\mathbb{R}^n$ is closed under scalar multiplication.)

(m.2) $1\mathbf{v} = \mathbf{v}$.

(m.3) $\alpha(\beta\mathbf{v}) = (\alpha\beta)\mathbf{v}$.

(m.4) $(\alpha + \beta)\mathbf{v} = \alpha\mathbf{v} + \beta\mathbf{v}$.

(m.5) $\alpha(\mathbf{v} + \mathbf{w}) = \alpha\mathbf{v} + \alpha\mathbf{w}$.

EXERCISE I

Check that all these properties are satisfied if $n = 1$, $n = 2$ or $n = 3$.

● Definition I

A set of vectors which satisfies the rules (a.1) to (a.5) and (m.1) to (m.5) is called a *real vector space*, and we refer to $\mathbb{R}^n$ as the *vector space* $\mathbb{R}^n$.

The reader should be aware that there are vector spaces which are different from any $\mathbb{R}^n$ (for example, the vector space of all functions mapping $\mathbb{R}$ to $\mathbb{R}$ which has infinite dimension), and although these are not within the scope of *this* book, they may be found in the book on linear algebra in this series.

5.2 Subspaces of $\mathbb{R}^n$

It is useful to know whether the vector space $\mathbb{R}^n$ contains any smaller vector spaces within it.

● Definition 2

A *subspace* of the vector space $\mathbb{R}^n$ is a non-empty subset S of $\mathbb{R}^n$ which is itself a vector space.

Since this book is primarily concerned with geometry rather than linear algebra, many of the theorems in this chapter will be quoted without a full proof. Proofs of these theorems will be found in any book on elementary linear algebra, but the reader is encouraged to check the properties as an exercise.

● *Theorem 1*

If S is a non-empty subset of the real vector space $\mathbb{R}^n$, then it is a *subspace* of $\mathbb{R}^n$ if and only if

$$\forall \mathbf{u}, \mathbf{v} \in S, \quad \mathbf{u} + \mathbf{v} \in S, \text{ and} \tag{5.2.1}$$

$$\forall \mathbf{u} \in S \text{ and } \forall \alpha \in \mathbb{R}, \quad \alpha \mathbf{u} \in S. \tag{5.2.2}$$

This means that S is closed under the operations of *vector addition* and *scalar multiplication*, and with these satisfied, the rest of the conditions for a vector space are also satisfied.

EXERCISE 2

Check that if a non-empty subset of $\mathbb{R}^3$ satisfies the conditions of Theorem 1, then it satisfies all the conditions for a vector space.

There is a one-step method for checking whether or not a subset is a subspace, and sometimes, though not always, this is a more convenient method.

● *Theorem 2*

If S is a non-empty subset of a real vector space V, then it is a *subspace* of V if and only if

$$\forall \mathbf{u}, \mathbf{v} \in S, \text{ and } \forall \alpha, \beta \in \mathbb{R}, \quad \alpha \mathbf{u} + \beta \mathbf{v} \in S. \tag{5.2.3}$$

● *Example 1*

Give a geometric description for each of the following subsets of $\mathbb{R}^3$, and, in each case, determine whether or not the subset is a subspace of $\mathbb{R}^3$.

(i) $S_1 = \left\{ \begin{pmatrix} x \\ y \\ z \end{pmatrix} \in \mathbb{R}^3 : x + y - z = 0 \right\}$,

(ii) $S_2 = \left\{ \begin{pmatrix} x \\ y \\ z \end{pmatrix} \in \mathbb{R}^3 : x + y - z = 1 \right\}$,

(iii) $S_3 = \left\{ \begin{pmatrix} x \\ y \\ z \end{pmatrix} \in \mathbb{R}^3 : x^2 + y^2 + z^2 + 1 = 0 \right\}$.

SOLUTION

(i) S_1 is a plane through the origin in $\mathbb{R}^3$. It is a non-empty set since $\begin{pmatrix} 0 \\ 0 \\ 0 \end{pmatrix} \in S_1$. Suppose $\mathbf{x}_1 = \begin{pmatrix} x_1 \\ y_1 \\ z_1 \end{pmatrix}$, $\mathbf{x}_2 = \begin{pmatrix} x_2 \\ y_2 \\ z_2 \end{pmatrix} \in S_1$ and $\alpha,\ \beta \in \mathbb{R}$.

Then $x_1 + y_1 - z_1 = 0$, and $x_2 + y_2 - z_2 = 0$. Now

$$\alpha \mathbf{x}_1 + \beta \mathbf{x}_2 = \begin{pmatrix} \alpha x_1 + \beta x_2 \\ \alpha y_1 + \beta y_2 \\ \alpha z_1 + \beta z_2 \end{pmatrix}$$

and

$$(\alpha x_1 + \beta x_2) + (\alpha y_1 + \beta y_2) - (\alpha z_1 + \beta z_2)$$
$$= \alpha(x_1 + y_1 - z_1) + \beta(x_2 + y_2 - z_2) = 0 + 0 = 0$$

Hence $\alpha \mathbf{x}_1 + \beta \mathbf{x}_2 \in S_1$, and so S_1 is a subspace of $\mathbb{R}^3$.

(ii) S_2 is a plane passing through the point $(1, 0, 0)$, but not through the origin. If $\mathbf{x} = \begin{pmatrix} x \\ y \\ z \end{pmatrix} \in S_2$, then $x + y - z = 1$. Now $2\mathbf{x} = \begin{pmatrix} 2x \\ 2y \\ 2z \end{pmatrix}$ and $(2x) + (2y) - (2z) = 2 \neq 1$. This means that $2\mathbf{x} \notin S_2$, so S_2 is not closed under scalar multiplication, and is therefore not a subspace of $\mathbb{R}^3$.

(iii) S_3 is the empty set, since there are no real numbers x, y, z such that $x^2 + y^2 + z^2 + 1 = 0$. A subspace of $\mathbb{R}^3$ must be a non-empty subset of $\mathbb{R}^3$, so S_3 is not a subspace of $\mathbb{R}^3$.

EXERCISE 3

Give a geometric description for each of the following subsets of $\mathbb{R}^3$, and, in each case, determine whether or not the subset is a subspace of $\mathbb{R}^3$.

(i) $S_4 = \left\{ \begin{pmatrix} x \\ y \\ z \end{pmatrix} \in \mathbb{R}^3 : x^2 + y^2 + z^2 = 0 \right\}$,

(ii) $S_5 = \left\{ \begin{pmatrix} x \\ y \\ z \end{pmatrix} \in \mathbb{R}^3 : x^2 + y^2 + z^2 = 1 \right\}$,

(iii) $S_6 = \left\{ \begin{pmatrix} x \\ y \\ z \end{pmatrix} \in \mathbb{R}^3 : x = y = z \right\}$.

So what are the subspaces of $\mathbb{R}^3$, for example? Any book on linear algebra will tell us that:

the only subspaces of $\mathbb{R}^3$ are (i) $\mathbb{R}^3$ itself, (ii) any plane passing through the origin, (iii) any line passing through the origin, (iv) the single point at the origin.

Similarly, the only subspaces of $\mathbb{R}^2$ are (i) $\mathbb{R}^2$ itself, (ii) any line passing through the origin, (iii) the single point at the origin.

Analogous results hold for subspaces of $\mathbb{R}^n$. (Again we see that although a *plane* is generally considered to be a set of *points*, in this case we define a plane, for example, by the set of all position vectors of points lying within it. Again there is an obvious equivalence.)

5.3 Linear combinations

We now need to consider how we can define subspaces of $\mathbb{R}^n$ in terms of vectors.

● *Definition 3*

Given the vectors $\mathbf{u}_1, \mathbf{u}_2, \ldots, \mathbf{u}_n$, we say that any vector $\mathbf{v}$ which can be written in the form

$$\mathbf{v} = \alpha_1\mathbf{u}_1 + \alpha_2\mathbf{u}_2 + \ldots + \alpha_n\mathbf{u}_n,$$

where $\alpha_1, \alpha_2, \ldots, \alpha_n \in \mathbb{R}$, is a *linear combination* of the vectors $\mathbf{u}_1, \mathbf{u}_2, \ldots, \mathbf{u}_n$.

● *Definition 4*

If every vector in a vector space V can be written as a linear combination of the vectors $\mathbf{v}_1, \ldots, \mathbf{v}_n$ of V, then we say that the vectors $\mathbf{v}_1, \ldots, \mathbf{v}_n$ *span* V, or that the set of vectors $\{\mathbf{v}_1, \ldots, \mathbf{v}_n\}$ is a *spanning set* for V.

● *Example 2*

Show that if

$$\mathbf{u} = \begin{pmatrix} 1 \\ 2 \\ 2 \end{pmatrix}, \quad \mathbf{v} = \begin{pmatrix} 2 \\ 1 \\ 1 \end{pmatrix}, \quad \mathbf{w} = \begin{pmatrix} 0 \\ 1 \\ 0 \end{pmatrix}, \quad \mathbf{t} = \begin{pmatrix} 1 \\ 0 \\ 0 \end{pmatrix},$$

(i) $\{\mathbf{u}, \mathbf{v}, \mathbf{w}\}$ is a spanning set for $\mathbb{R}^3$, but
(ii) $\{\mathbf{u}, \mathbf{v}, \mathbf{t}\}$ is not a spanning set for $\mathbb{R}^3$.

SOLUTION

(i) Suppose that $\mathbf{x} = \begin{pmatrix} x \\ y \\ z \end{pmatrix}$ is a general element of $\mathbb{R}^3$, and that

$$\mathbf{x} = \alpha\begin{pmatrix} 1 \\ 2 \\ 2 \end{pmatrix} + \beta\begin{pmatrix} 2 \\ 1 \\ 1 \end{pmatrix} + \gamma\begin{pmatrix} 0 \\ 1 \\ 0 \end{pmatrix}.$$

Then

$$x = \alpha + 2\beta \tag{1}$$
$$y = 2\alpha + \beta + \gamma \tag{2}$$
$$z = 2\alpha + \beta. \tag{3}$$

Multiplying (3) by 2 and subtracting it from (1) gives

$$x - 2z = \alpha - 4\alpha \quad \Rightarrow \quad \alpha = \frac{1}{3}(2z - x)$$

and by back substitution this leads to

$$\beta = \frac{1}{3}(2x - z) \quad \text{and} \quad \gamma = y - z.$$

Hence

$$\begin{pmatrix} x \\ y \\ z \end{pmatrix} = \frac{1}{3}(2z - x)\begin{pmatrix} 1 \\ 2 \\ 2 \end{pmatrix} + \frac{1}{3}(2x - z)\begin{pmatrix} 2 \\ 1 \\ 1 \end{pmatrix} + (y - z)\begin{pmatrix} 0 \\ 1 \\ 0 \end{pmatrix}.$$

(As usual, it is sensible to check that this result is correct by checking the components one at a time.) This shows that *any* element of $\mathbb{R}^3$ can be written as a linear combination of $\mathbf{u}, \mathbf{v}$ and $\mathbf{w}$ and so $\{\mathbf{u}, \mathbf{v}, \mathbf{w}\}$ is a spanning set for $\mathbb{R}^3$.

(ii)   Suppose that

$$\mathbf{x} = \alpha\begin{pmatrix} 1 \\ 2 \\ 2 \end{pmatrix} + \beta\begin{pmatrix} 2 \\ 1 \\ 1 \end{pmatrix} + \gamma\begin{pmatrix} 1 \\ 0 \\ 0 \end{pmatrix}.$$

Then

$$x = \alpha + 2\beta + \gamma \tag{1}$$
$$y = 2\alpha + \beta \tag{2}$$
$$z = 2\alpha + \beta. \tag{3}$$

From this we see that $y = z$, and so there would be no solution in α, β, γ for any vector for which the y and z components are different. For example, the vector $\begin{pmatrix} 2 \\ 1 \\ 3 \end{pmatrix}$ in $\mathbb{R}^3$ cannot be expressed as a linear combination of $\mathbf{u}, \mathbf{v}, \mathbf{t}$, and hence $\{\mathbf{u}, \mathbf{v}, \mathbf{t}\}$ is not a spanning set for $\mathbb{R}^3$.

EXERCISE 4

Determine whether or not the set of vectors $\{\mathbf{u}, \mathbf{v}, \mathbf{w}\}$ is a spanning set for $\mathbb{R}^3$ in each of the following cases:

(i) $\mathbf{u} = \begin{pmatrix} 2 \\ 1 \\ 1 \end{pmatrix}$, $\mathbf{v} = \begin{pmatrix} 3 \\ -2 \\ -2 \end{pmatrix}$, $\mathbf{w} = \begin{pmatrix} 4 \\ 4 \\ 4 \end{pmatrix}$

(ii) $\mathbf{u} = \begin{pmatrix} 2 \\ 1 \\ 0 \end{pmatrix}$, $\mathbf{v} = \begin{pmatrix} 0 \\ 1 \\ 0 \end{pmatrix}$, $\mathbf{w} = \begin{pmatrix} 1 \\ 0 \\ 2 \end{pmatrix}$.

In Chapter 1 we saw that any vector in $\mathbb{R}^3$ can be written as a linear combination of the vectors $\mathbf{i}, \mathbf{j}$ and $\mathbf{k}$. Hence $\{\mathbf{i}, \mathbf{j}, \mathbf{k}\}$ is a spanning set for $\mathbb{R}^3$.

● Definition 5

The vectors $\mathbf{v}_1, \mathbf{v}_2, \ldots, \mathbf{v}_n$ are said to be *linearly dependent* if we can find scalars $\alpha_1, \alpha_2, \ldots, \alpha_n$ not all zero, such that

$$\alpha_1\mathbf{v}_1 + \alpha_2\mathbf{v}_2 + \ldots + \alpha_n\mathbf{v}_n = \mathbf{0}.$$

Remember that it is the zero *vector* on the right.

If the vectors $v_1, v_2, \ldots, v_n$ are not linearly dependent, then they are said to be *linearly independent*, but we often use the following equivalent definition.

● Definition 6

The vectors $v_1, v_2, \ldots, v_n$ are said to be *linearly independent* if for scalars $\alpha_1, \alpha_2, \ldots, \alpha_n$,

$$\alpha_1 v_1 + \alpha_2 v_2 + \ldots + \alpha_n v_n = 0$$

implies that $\alpha_1 = \alpha_2 = \ldots = \alpha_n = 0$.

● Definition 7

If $A = \{v_1, v_2, \ldots, v_n\}$ is a set of vectors which are linearly independent, then we say A is a *linearly independent set*, otherwise we say A is a *linearly dependent set*.

● Example 3

Show whether the set of vectors $\{u, v, w\}$ is linearly dependent or linearly independent in each of the following cases:

(i) $u = \begin{pmatrix} 2 \\ 1 \\ -3 \end{pmatrix}$, $v = \begin{pmatrix} 1 \\ 0 \\ 2 \end{pmatrix}$, $w = \begin{pmatrix} 1 \\ 1 \\ 1 \end{pmatrix}$.

(ii) $u = \begin{pmatrix} 1 \\ 2 \\ 3 \end{pmatrix}$, $v = \begin{pmatrix} 2 \\ 1 \\ 2 \end{pmatrix}$, $w = \begin{pmatrix} 0 \\ -3 \\ -4 \end{pmatrix}$.

SOLUTION

(i) Suppose $\alpha u + \beta v + \gamma w = 0$; then

$$2\alpha + \beta + \gamma = 0$$
$$\alpha + \gamma = 0$$
$$-3\alpha + 2\beta + \gamma = 0.$$

From the second equation $\alpha = -\gamma$ and by substituting this into the other two equations we get

$$\alpha + \beta = 0 \quad \text{and} \quad -4\alpha + 2\beta = 0.$$

The only solution to these equations is then $\alpha = \beta = \gamma = 0$, which means that the three vectors are linearly independent.

(ii) Suppose $\alpha u + \beta v + \gamma w = 0$; then

$$\alpha + 2\beta = 0$$
$$2\alpha + \beta - 3\gamma = 0$$
$$3\alpha + 2\beta - 4\gamma = 0.$$

From the first equation $\alpha = -2\beta$, and substituting into the other equations gives

$$-3\beta - 3\gamma = 0 \quad \text{and} \quad -4\beta - 4\gamma = 0.$$

So if $\gamma = 1$, $\beta = -1$ and $\alpha = 2$, all three equations are satisfied, and since α, β, γ are not all zero, the three vectors are linearly dependent.

EXERCISE 5

In each of the following cases determine whether or not the vectors (a) are linearly independent, or (b) form a spanning set for $\mathbb{R}^3$.

(i) $\mathbf{u} = \begin{pmatrix} 1 \\ 3 \\ 2 \end{pmatrix}$, $\mathbf{v} = \begin{pmatrix} -2 \\ 1 \\ 1 \end{pmatrix}$, $\mathbf{w} = \begin{pmatrix} 7 \\ 7 \\ 4 \end{pmatrix}$.

(ii) $\mathbf{u} = \begin{pmatrix} 1 \\ 2 \\ 1 \end{pmatrix}$, $\mathbf{v} = \begin{pmatrix} 4 \\ 4 \\ 2 \end{pmatrix}$, $\mathbf{w} = \begin{pmatrix} 2 \\ 0 \\ 3 \end{pmatrix}$.

(iii) $\mathbf{u} = \begin{pmatrix} 1 \\ 0 \\ 1 \end{pmatrix}$, $\mathbf{v} = \begin{pmatrix} 1 \\ 2 \\ 1 \end{pmatrix}$, $\mathbf{w} = \begin{pmatrix} 1 \\ 0 \\ 3 \end{pmatrix}$, $\mathbf{x} = \begin{pmatrix} 3 \\ 0 \\ 1 \end{pmatrix}$.

(iv) $\mathbf{u} = \begin{pmatrix} 2 \\ 0 \\ 1 \end{pmatrix}$, $\mathbf{v} = \begin{pmatrix} 2 \\ 2 \\ 1 \end{pmatrix}$, $\mathbf{w} = \begin{pmatrix} 0 \\ 1 \\ 0 \end{pmatrix}$, $\mathbf{x} = \begin{pmatrix} 4 \\ 3 \\ 2 \end{pmatrix}$.

Let us consider the geometrical significance of these definitions. If we have a single vector $\mathbf{x}$ in $\mathbb{R}^3$, then unless it is the zero vector, we can only have $\alpha\mathbf{x} = \mathbf{0}$ when $\alpha = 0$. This means that a single non-zero vector is linearly independent. On the other hand if we have a set of vectors $\{\mathbf{v}_1, \ldots, \mathbf{v}_n\}$ where $\mathbf{v}_i = \mathbf{0}$ for some i, $1 \le i \le n$, then

$$\alpha_1\mathbf{v}_1 + \ldots + \alpha_n\mathbf{v}_n = \mathbf{0}$$

is satisfied when $\alpha_i = 1$ and $\alpha_j = 0$, $j \ne i$. So not all the α_k $(k = 1, \ldots, n)$ are zero, and the vectors are linearly dependent. This means that *any set of vectors containing the zero vector is a linearly dependent set*.

Suppose the vectors $\mathbf{v}_1, \mathbf{v}_2, \ldots, \mathbf{v}_n$ are linearly dependent; then, from the definition, there exist real numbers $\alpha_1, \alpha_2, \ldots, \alpha_n$ not all zero such that

$$\alpha_1\mathbf{v}_1 + \alpha_2\mathbf{v}_2 + \ldots + \alpha_n\mathbf{v}_n = \mathbf{0}$$

and we can suppose without loss of generality that $\alpha_1 \ne 0$ (for we can simply relabel the vectors so that this is true). Then

$$\mathbf{v}_1 = -\frac{\alpha_2}{\alpha_1}\mathbf{v}_2 - \ldots - \frac{\alpha_n}{\alpha_1}\mathbf{v}_n.$$

Now if $n = 2$ this means that $\mathbf{v}_1$ is a multiple of $\mathbf{v}_2$ which means that $\mathbf{v}_1$ is parallel to $\mathbf{v}_2$. Similarly if $n = 3$ then $\mathbf{v}_1$ is a linear combination of $\mathbf{v}_2$ and $\mathbf{v}_3$ and hence $\mathbf{v}_1$ is coplanar with the vectors $\mathbf{v}_2$ and $\mathbf{v}_3$. This means that any two non-zero vectors in $\mathbb{R}^3$ are linearly independent if they are not parallel, and any three non-zero vectors are linearly independent if they are not coplanar. In fact this is true in $\mathbb{R}^n$ for any n, but if $n = 2$ all vectors in $\mathbb{R}^2$ are coplanar, so there can be no more than two linearly independent vectors in $\mathbb{R}^2$. In an analogous way we can have at most three linearly independent vectors in $\mathbb{R}^3$, and at most n linearly independent vectors in $\mathbb{R}^n$.

Bearing these facts in mind we can see that any spanning set for $\mathbb{R}^3$ must contain *at least* three vectors, for if $\mathbf{u}$ and $\mathbf{v}$ are any two non-zero, non-parallel vectors in $\mathbb{R}^3$, then $\mathbf{u} \times \mathbf{v}$ is also a non-zero vector of $\mathbb{R}^3$. Suppose

$$\mathbf{u} \times \mathbf{v} = \alpha\mathbf{u} + \beta\mathbf{v};$$

then, taking the scalar product of $\mathbf{u} \times \mathbf{v}$ (which is perpendicular to both $\mathbf{u}$ and $\mathbf{v}$), with each side of this equation, we get

$$(\mathbf{u} \times \mathbf{v}).(\mathbf{u} \times \mathbf{v}) = (\mathbf{u} \times \mathbf{v}).\alpha\mathbf{u} + (\mathbf{u} \times \mathbf{v}).\beta\mathbf{v} = 0.$$

But this is a contradiction, since $(\mathbf{u} \times \mathbf{v}).(\mathbf{u} \times \mathbf{v}) \neq 0$ since $\mathbf{u} \times \mathbf{v}$ is a non-zero vector. Hence we have found a vector, namely $\mathbf{u} \times \mathbf{v}$ in $\mathbb{R}^3$, which cannot be a linear combination of $\mathbf{u}$ and $\mathbf{v}$, so at least three vectors are needed to span $\mathbb{R}^3$. Similarly, a spanning set of $\mathbb{R}^2$ contains at least two vectors, and, more generally, a spanning set of $\mathbb{R}^n$ contains at least n vectors.

Note that a spanning set of $\mathbb{R}^3$ can be found which contains *more than* three vectors. Take for example $\{\mathbf{i}, \mathbf{j}, \mathbf{k}, \mathbf{i} + \mathbf{j}\}$. This is a spanning set for $\mathbb{R}^3$ containing four vectors, but note that it is not a linearly independent set. Likewise $\{\mathbf{i}, \mathbf{j}\}$ is a linearly independent set of vectors from $\mathbb{R}^3$, but it is not a spanning set for $\mathbb{R}^3$.

5.4 Bases for vector spaces

We have already seen that $\mathbf{i}$, $\mathbf{j}$ and $\mathbf{k}$ are important to the structure of $\mathbb{R}^3$ in that they are linearly independent vectors which span $\mathbb{R}^3$; that is, any vector in $\mathbb{R}^3$ can be written as a linear combination of these three vectors, and more than that, the linear combination for each vector of $\mathbb{R}^3$ is unique. However, they are not the only triad of vectors for which this is true, and the next definition will give us a more general case.

● *Definition 8*

In a vector space V the subset $B = \{\mathbf{v}_1, \mathbf{v}_2, \ldots, \mathbf{v}_n\}$ is a *basis* for V if

 (i) the vectors of B are linearly independent, and
(ii) the vectors of B span V.

● *Definition 9*

We call $\{\mathbf{i}, \mathbf{j}, \mathbf{k}\}$ the *standard basis* for $\mathbb{R}^3$.

● *Theorem 3* ———————————————————————

A basis for $\mathbb{R}^n$ contains exactly n vectors.

Although no rigorous proof is given here, from the geometric arguments at the end of Section 5.3, we saw that a linearly independent set in $\mathbb{R}^n$ contains at most n vectors, and a spanning set for $\mathbb{R}^n$ contains at least n vectors. Hence if B is a basis for $\mathbb{R}^n$ it must contain at least n vectors, and at most n vectors; that is, it must contain *exactly n* vectors.

● *Theorem 4* ───────────────────────────────────

(i) If $B = \{v_1, v_2, \ldots, v_n\}$ is a subset of $\mathbb{R}^n$ comprising n linearly independent vectors, then B is a basis for $\mathbb{R}^n$.
(ii) If S is a subspace of $\mathbb{R}^n$, and if $B' = \{v_1, v_2, \ldots, v_k\}$ is a basis of S containing k vectors, then any set of k linearly independent vectors of S is a basis for S.

● *Definition 10*

We say that a subspace S of $\mathbb{R}^n$ has dimension k when there are k vectors in a basis for S, and hence that $\mathbb{R}^n$ has *dimension n*.

This means that the above definition of *dimension* agrees with our usual notion of dimension, so that a plane has dimension 2, a line has dimension 1, and what we think of as three-dimensional space has dimension 3. A point is zero-dimensional.

● *Example 4*

Show whether or not the following subsets of $\mathbb{R}^3$ are bases of $\mathbb{R}^3$:

(i) $S_1 = \left\{ \begin{pmatrix} 1 \\ 2 \\ 1 \end{pmatrix}, \begin{pmatrix} 1 \\ 2 \\ 3 \end{pmatrix}, \begin{pmatrix} 2 \\ -1 \\ 2 \end{pmatrix} \right\}$, (ii) $S_2 = \left\{ \begin{pmatrix} 1 \\ -1 \\ 0 \end{pmatrix}, \begin{pmatrix} 0 \\ 1 \\ -1 \end{pmatrix}, \begin{pmatrix} -1 \\ 0 \\ 1 \end{pmatrix} \right\}$.

SOLUTION

Suppose $\alpha \begin{pmatrix} 1 \\ 2 \\ 1 \end{pmatrix} + \beta \begin{pmatrix} 1 \\ 2 \\ 3 \end{pmatrix} + \gamma \begin{pmatrix} 2 \\ -1 \\ 2 \end{pmatrix} = \begin{pmatrix} 0 \\ 0 \\ 0 \end{pmatrix}$; then

$$\alpha + \beta + 2\gamma = 0, \tag{I}$$
$$2\alpha + 2\beta - \gamma = 0, \tag{II}$$
$$\alpha + 3\beta + 2\gamma = 0. \tag{III}$$

Adding $2 \times$ (II) first to (I) and then to (III) we get

$$5\alpha + 5\beta = 0 \quad \text{and} \quad 5\alpha + 7\beta = 0$$

and the only solution to these, and hence to all three equations, is $\alpha = \beta = \gamma = 0$, which means that the three vectors are linearly independent.

From Theorem 4, since we have three linearly independent vectors of $\mathbb{R}^3$, we have a basis for $\mathbb{R}^3$.

(ii) Since

$$\begin{pmatrix} 1 \\ -1 \\ 0 \end{pmatrix} + \begin{pmatrix} 0 \\ 1 \\ -1 \end{pmatrix} + \begin{pmatrix} -1 \\ 0 \\ 1 \end{pmatrix} = \begin{pmatrix} 0 \\ 0 \\ 0 \end{pmatrix},$$

the three vectors are linearly dependent and so cannot form a basis for $\mathbb{R}^3$.

EXERCISE 6

Show whether or not the following subsets of $\mathbb{R}^3$ are bases of $\mathbb{R}^3$:

(i) $S_1 = \left\{ \begin{pmatrix} 1 \\ 1 \\ 2 \end{pmatrix}, \begin{pmatrix} 2 \\ -1 \\ 1 \end{pmatrix}, \begin{pmatrix} -1 \\ 5 \\ 4 \end{pmatrix} \right\}$,   (ii) $S_2 = \left\{ \begin{pmatrix} 1 \\ 1 \\ 2 \end{pmatrix}, \begin{pmatrix} 0 \\ 1 \\ -1 \end{pmatrix}, \begin{pmatrix} 2 \\ 0 \\ 1 \end{pmatrix} \right\}$.

5.5 Orthogonal bases

• Definition 11

A basis B of $\mathbb{R}^n$ is an *orthogonal basis* if the vectors of B are mutually orthogonal (that is, any two distinct vectors of B are orthogonal).

Suppose $B = \{v_1, v_2, v_3\}$ is an orthogonal basis of $\mathbb{R}^3$. Then

$$v_1 \cdot v_2 = v_2 \cdot v_3 = v_3 \cdot v_1 = 0. \tag{5.5.1}$$

In this case it is much easier to express a general vector of $\mathbb{R}^3$ as a linear combination of these basis vectors. For suppose

$$x = \alpha v_1 + \beta v_2 + \gamma v_3.$$

Then taking the scalar product of each side of this equation with v_1, and using equations (5.5.1), we get

$$x \cdot v_1 = \alpha v_1 \cdot v_1 + \beta v_2 \cdot v_1 + \gamma v_3 \cdot v_1 = \alpha v_1 \cdot v_1.$$

Now $v_1 \cdot v_1 \neq 0$ since no basis vector can be the zero vector, and $v_1 \cdot v_2 = v_1 \cdot v_3 = 0$ from equation (5.5.1), so

$$\alpha = \frac{x \cdot v_1}{v_1 \cdot v_1}.$$

In this way α is calculated without having to solve simultaneous equations, and by similar means taking the scalar product each side of the equation with v_2 and v_3, we find that

$$\beta = \frac{x \cdot v_2}{v_2 \cdot v_2}, \quad \gamma = \frac{x \cdot v_3}{v_3 \cdot v_3}.$$

EXERCISE 7

Show that $\{v_1, v_2, v_3\}$ is an orthogonal basis for $\mathbb{R}^3$, and find the vector $x = \begin{pmatrix} 2 \\ 1 \\ 4 \end{pmatrix}$ as a linear combination of these basis vectors when

$$v_1 = \begin{pmatrix} 1 \\ -1 \\ 1 \end{pmatrix}, \quad v_2 = \begin{pmatrix} 1 \\ 0 \\ -1 \end{pmatrix}, \quad v_3 = \begin{pmatrix} 1 \\ 2 \\ 1 \end{pmatrix}.$$

(*Hint*: After proving that the three vectors do form an orthogonal basis (that is, they form a basis and are mutually orthogonal), suppose

$$x = \alpha v_1 + \beta v_2 + \gamma v_3.$$

Then

$$\alpha = \frac{\mathbf{x} \cdot \mathbf{v}_1}{\mathbf{v}_1 \cdot \mathbf{v}_1}.$$

Evaluate this and find β and γ in a similar way.)

This is all very well, but our basis may not be orthogonal to begin with, and we might wish to keep a particular vector as part of the basis, and to find an orthogonal basis containing this vector. Alternatively, we might be in a three-dimensional subspace of $\mathbb{R}^n$ where $n > 3$. It would be convenient to be able to find an orthogonal basis for any subspace of $\mathbb{R}^n$, and this is what the next section is all about.

5.6 Gram-Schmidt orthogonalisation process

Suppose we have a three-dimensional subspace S of $\mathbb{R}^n$ with basis $\{\mathbf{v}_1, \mathbf{v}_2, \mathbf{v}_3\}$. Then we set

(i) $\mathbf{u}_1 = \mathbf{v}_1$

(ii) $\lambda \mathbf{u}_2 = \mathbf{v}_2 - \dfrac{\mathbf{v}_2 \cdot \mathbf{u}_1}{\mathbf{u}_1 \cdot \mathbf{u}_1} \mathbf{u}_1$

(iii) $\mu \mathbf{u}_3 = \mathbf{v}_3 - \dfrac{\mathbf{v}_3 \cdot \mathbf{u}_1}{\mathbf{u}_1 \cdot \mathbf{u}_1} \mathbf{u}_1 - \dfrac{\mathbf{v}_3 \cdot \mathbf{u}_2}{\mathbf{u}_2 \cdot \mathbf{u}_2} \mathbf{u}_2.$

Then $\{\mathbf{u}_1, \mathbf{u}_2, \mathbf{u}_3\}$ will be an orthogonal basis for S. The λ and μ are non-zero real numbers chosen so that the coordinates of $\mathbf{u}_2$ and $\mathbf{u}_3$ are as convenient as possible. (See the example following Theorem 5.)

Why is this? First we show that the vectors $\mathbf{u}_1, \mathbf{u}_2, \mathbf{u}_3$ are orthogonal, noting that, since λ is non-zero, $\lambda \mathbf{u}_2$ is orthogonal to a vector $\mathbf{w}$ if and only if $\mathbf{u}_2$ is orthogonal to $\mathbf{w}$.

$$\lambda \mathbf{u}_2 \cdot \mathbf{u}_1 = \mathbf{v}_2 \cdot \mathbf{u}_1 - \frac{\mathbf{v}_2 \cdot \mathbf{u}_1}{\mathbf{u}_1 \cdot \mathbf{u}_1} \mathbf{u}_1 \cdot \mathbf{u}_1 = \mathbf{v}_2 \cdot \mathbf{u}_1 - \mathbf{v}_2 \cdot \mathbf{u}_1 = 0$$

that is

$$\mathbf{u}_1 \cdot \mathbf{u}_2 = 0 \qquad\qquad (5.6.1)$$

which means $\mathbf{u}_2$ is orthogonal to $\mathbf{u}_1$.

EXERCISE 8

Show first that $\mathbf{u}_1$ is orthogonal to $\mathbf{u}_3$, and next that $\mathbf{u}_2$ is orthogonal to $\mathbf{u}_3$. (*Hint:* You will need to use the first of these results when proving the second.)

To show that these vectors form a basis for S we need the following theorem.

• *Theorem 5*——————————————

If $\mathbf{v}_1, \mathbf{v}_2, \ldots \mathbf{v}_n$ are mutually orthogonal non-zero vectors, then they are linearly independent.

PROOF

Suppose $\alpha_1, \alpha_2, \ldots, \alpha_n$ are real numbers such that

$$\alpha_1 \mathbf{v}_1 + \alpha_2 \mathbf{v}_2 + \ldots + \alpha_n \mathbf{v}_n = \mathbf{0}.$$

Then by taking the scalar product of each side with $\mathbf{v}_1$ we see that

$$\alpha_1 \mathbf{v}_1 . \mathbf{v}_1 + \alpha_2 \mathbf{v}_2 . \mathbf{v}_1 + \ldots + \alpha_n \mathbf{v}_n . \mathbf{v}_1 = \mathbf{0} . \mathbf{v}_1.$$

But because of the orthogonality ($\mathbf{v}_i . \mathbf{v}_1 = 0$ if $i \neq 1$) this reduces to

$$\alpha_1 \mathbf{v}_1 . \mathbf{v}_1 = 0$$

and since $\mathbf{v}_1 \neq \mathbf{0}$ this implies that $\alpha_1 = 0$. Similarly we can take the scalar product of each side of the equation with $\mathbf{v}_2, \ldots, \mathbf{v}_n$ in turn to show that $\alpha_2 = \ldots = \alpha_n = 0$, and hence the vectors are linearly independent.

Now $\mathbf{u}_1, \mathbf{u}_2, \mathbf{u}_3$ are orthogonal, so by Theorem 5 they are linearly independent, and since each of the vectors is a linear combination of vectors in S, it is itself a vector in S. Since S has dimension 3, as there were three vectors in the original basis, any three linearly independent vectors of S form a basis for S, so it follows that $\{\mathbf{u}_1, \mathbf{u}_2, \mathbf{u}_3\}$ is an orthogonal basis for S.

Example 5

Use the Gram-Schmidt orthogonalisation process to obtain three orthogonal vectors, given the vectors

$$\mathbf{v}_1 = \begin{pmatrix} 1 \\ 2 \\ 1 \end{pmatrix}, \qquad \mathbf{v}_2 = \begin{pmatrix} 1 \\ 1 \\ -1 \end{pmatrix}, \qquad \mathbf{v}_3 = \begin{pmatrix} 2 \\ -1 \\ 1 \end{pmatrix}.$$

SOLUTION

Set $\mathbf{u}_1 = \mathbf{v}_1$; then

$$\lambda \mathbf{u}_2 = \mathbf{v}_2 - \frac{\mathbf{v}_2 . \mathbf{u}_1}{\mathbf{u}_1 . \mathbf{u}_1} \mathbf{u}_1$$

$$= \begin{pmatrix} 1 \\ 1 \\ -1 \end{pmatrix} - \frac{2}{6} \begin{pmatrix} 1 \\ 2 \\ 1 \end{pmatrix}$$

$$= \frac{1}{3} \begin{pmatrix} 2 \\ 1 \\ -4 \end{pmatrix}.$$

By choosing $\lambda = 1/3$, we have $\mathbf{u}_2 = \begin{pmatrix} 2 \\ 1 \\ -4 \end{pmatrix}$. Notice how much more convenient it is to have no fractions in the vector, and it is even more convenient at the next stage. This is the reason for putting the factor λ in.

$$\mu \mathbf{u}_3 = \mathbf{v}_3 - \frac{\mathbf{v}_3 \cdot \mathbf{u}_1}{\mathbf{u}_1 \cdot \mathbf{u}_1} \mathbf{u}_1 - \frac{\mathbf{v}_3 \cdot \mathbf{u}_2}{\mathbf{u}_2 \cdot \mathbf{u}_2} \mathbf{u}_2$$

$$= \begin{pmatrix} 2 \\ -1 \\ 1 \end{pmatrix} - \frac{1}{6} \begin{pmatrix} 1 \\ 2 \\ 1 \end{pmatrix} + \frac{1}{21} \begin{pmatrix} 2 \\ 1 \\ -4 \end{pmatrix}$$

$$= \frac{27}{42} \begin{pmatrix} 3 \\ -2 \\ 1 \end{pmatrix}.$$

Hence, by choosing $\mu = 27/42$ our three orthogonal vectors are

$$\mathbf{u}_1 = \begin{pmatrix} 1 \\ 2 \\ 1 \end{pmatrix}, \qquad \mathbf{u}_2 = \begin{pmatrix} 2 \\ 1 \\ -4 \end{pmatrix}, \qquad \mathbf{u}_3 = \begin{pmatrix} 3 \\ -2 \\ 1 \end{pmatrix}.$$

Check that these vectors really are orthogonal.

EXERCISE 9

Suppose $\{\mathbf{v}_1, \mathbf{v}_2, \mathbf{v}_3\}$ forms a basis for S, a subspace of $\mathbb{R}^4$, where

$$\mathbf{v}_1 = \begin{pmatrix} 1 \\ 1 \\ 1 \\ 0 \end{pmatrix}, \qquad \mathbf{v}_2 = \begin{pmatrix} 0 \\ 1 \\ 2 \\ 0 \end{pmatrix}, \qquad \mathbf{v}_3 = \begin{pmatrix} 1 \\ -1 \\ 0 \\ 2 \end{pmatrix}.$$

Find an orthogonal basis for S.

Because it is often convenient to have basis vectors which are unit vectors, the following definition is useful.

● Definition 12

An orthogonal basis, whose vectors are all unit vectors, is called an *orthonormal* basis.

It is easy to convert an orthogonal basis into an orthonormal basis by multiplying each vector by the reciprocal of its length.

EXERCISE 10

Convert the orthogonal basis you found in Exercise 9 into an orthonormal basis.

Summary

1. $\mathbb{R}^n$ satisfies the conditions for a vector space.
2. The subspaces of $\mathbb{R}^3$ are (i) $\mathbb{R}^3$ itself, (ii) any plane through the origin, (iii) any line through the origin, (iv) the origin itself. Similarly the subspaces of $\mathbb{R}^2$ are (i) $\mathbb{R}^2$ itself, (ii) any line through the origin, (iii) the origin itself.

3. We defined the terms *linear combination*, *linear dependence* and *linear indepen-dence* in definitions 5, 6 and 7 respectively.
4. A *basis* for a vector space V is a set S of vectors of V such that S is both a linearly independent set and a spanning set for V.
5. The *dimension* of a vector space V is the number of vectors in a basis of V, which is constant for V.
6. In an orthogonal basis, the vectors are mutually orthogonal, and the Gram-Schmidt process is a method for finding an orthogonal basis from a given basis.
7. A set of mutually orthogonal vectors is a linearly independent set of vectors.
8. An orthonormal basis for a vector space V is an orthogonal basis whose vectors are all unit vectors.

FURTHER EXERCISES

11. Suppose that

$$\mathbf{a} = \begin{pmatrix} 1 \\ 2 \\ 2 \\ 3 \end{pmatrix}, \quad \mathbf{b} = \begin{pmatrix} 0 \\ 1 \\ 2 \\ -2 \end{pmatrix}, \quad \mathbf{c} = \begin{pmatrix} 1 \\ 1 \\ 1 \\ 1 \end{pmatrix}, \quad \mathbf{d} = \begin{pmatrix} 1 \\ -3 \\ 1 \\ 1 \end{pmatrix}.$$

(i) Find the lengths of the vectors $\mathbf{a}$, $\mathbf{b}$, $\mathbf{c}$, $\mathbf{d}$.
(ii) Find the scalar products $\mathbf{a.b}$, $\mathbf{a.c}$, $\mathbf{a.d}$, $\mathbf{b.c}$, $\mathbf{b.d}$, $\mathbf{c.d}$.
(iii) Which pairs of vectors are orthogonal?

12. Describe each of the following subsets of $\mathbb{R}^3$ geometrically, and determine whether or not each is a subspace of $\mathbb{R}^3$.

(i) $S = \left\{ \begin{pmatrix} x \\ y \\ z \end{pmatrix} \in \mathbb{R}^3 : 2x = y \right\},$

(ii) $T = \left\{ \begin{pmatrix} x \\ y \\ z \end{pmatrix} \in \mathbb{R}^3 : y = z = 0 \right\},$

(iii) $U = \left\{ \begin{pmatrix} x \\ y \\ z \end{pmatrix} \in \mathbb{R}^3 : xyz = 0 \right\}.$

13. Given that $\mathbf{u} = \begin{pmatrix} 1 \\ 2 \\ 1 \end{pmatrix}$ and $\mathbf{v} = \begin{pmatrix} 2 \\ -1 \\ 2 \end{pmatrix}$, find whether or not each of the following vectors can be written as a linear combination of $\mathbf{u}$ and $\mathbf{v}$, and, where possible, write down this linear combination:

(i) $\mathbf{a} = \begin{pmatrix} 4 \\ 3 \\ 4 \end{pmatrix},$ (ii) $\mathbf{b} = \begin{pmatrix} 1 \\ 3 \\ 4 \end{pmatrix},$ (iii) $\mathbf{c} = \begin{pmatrix} 1 \\ 3 \\ 1 \end{pmatrix}.$

14. 'S is a basis of V'. In each of the following cases, determine whether or not this statement is true, giving reasons for your answer.

(i) $S = \left\{ \begin{pmatrix} 1 \\ 2 \end{pmatrix}, \begin{pmatrix} 2 \\ 1 \end{pmatrix} \right\}$, $V = \mathbb{R}^2$

(ii) $S = \left\{ \begin{pmatrix} 1 \\ 2 \end{pmatrix}, \begin{pmatrix} 2 \\ 1 \end{pmatrix}, \begin{pmatrix} 1 \\ 3 \end{pmatrix} \right\}$, $V = \mathbb{R}^2$

(iii) $S = \left\{ \begin{pmatrix} 1 \\ 2 \\ 1 \end{pmatrix}, \begin{pmatrix} 5 \\ 3 \\ 2 \end{pmatrix} \right\}$, $V = \mathbb{R}^3$

(iv) $S = \left\{ \begin{pmatrix} 1 \\ 2 \\ 1 \end{pmatrix}, \begin{pmatrix} 1 \\ 0 \\ 1 \end{pmatrix}, \begin{pmatrix} 1 \\ 0 \\ -1 \end{pmatrix} \right\}$, $V = \mathbb{R}^3$.

15. Given that $\mathbf{u}_1 = \begin{pmatrix} 1 \\ 2 \\ 3 \end{pmatrix}$, $\mathbf{u}_2 = \begin{pmatrix} 1 \\ 1 \\ -1 \end{pmatrix}$, $\mathbf{u}_3 = \begin{pmatrix} -5 \\ 4 \\ -1 \end{pmatrix}$ and $\mathbf{x} = \begin{pmatrix} 2 \\ 2 \\ 5 \end{pmatrix}$, show that

$\{\mathbf{u}_1, \mathbf{u}_2, \mathbf{u}_3\}$ is an orthogonal basis for $\mathbb{R}^3$, and express $\mathbf{x}$ as a linear combination of these basis vectors.

16. (i) Suppose $\mathbf{p} = \begin{pmatrix} 1 \\ 1 \\ 1 \end{pmatrix}$, $\mathbf{q} = \begin{pmatrix} -1 \\ 0 \\ 1 \end{pmatrix}$, $\mathbf{x} = \begin{pmatrix} a \\ b \\ c \end{pmatrix}$, and $\mathbf{x}$ is a unit vector. Given

that $\mathbf{p}.\mathbf{x} = 0$ and $\mathbf{q}.\mathbf{x} = 0$, find a, b and c. Explain why $\mathbf{p}$, $\mathbf{q}$ and $\mathbf{x}$ are mutually orthogonal.

(ii) With the $\mathbf{p}$ and $\mathbf{q}$ from (i), find $\mathbf{p} \times \mathbf{q}$. How does this compare with the $\mathbf{x}$ of (i) and why?

(iii) Let $\mathbf{v}_1 = \begin{pmatrix} 1 \\ 1 \\ 1 \end{pmatrix}$, $\mathbf{v}_2 = \begin{pmatrix} 1 \\ 2 \\ 3 \end{pmatrix}$ and $\mathbf{v}_3 = \begin{pmatrix} 1 \\ 3 \\ 2 \end{pmatrix}$. Use the Gram-Schmidt

orthogonalisation process to find three orthogonal vectors in $\mathbb{R}^3$ one of which is $\mathbf{v}_1$. Compare your answer with the results of (i) and (ii).

17. Suppose $\{\mathbf{v}_1, \mathbf{v}_2, \mathbf{v}_3\}$ forms a basis for S, a subspace of $\mathbb{R}^4$, where

$$\mathbf{v}_1 = \begin{pmatrix} 1 \\ 1 \\ 0 \\ 0 \end{pmatrix}, \qquad \mathbf{v}_2 = \begin{pmatrix} 0 \\ 1 \\ 1 \\ 0 \end{pmatrix}, \qquad \mathbf{v}_3 = \begin{pmatrix} 1 \\ 0 \\ 1 \\ 0 \end{pmatrix}.$$

Find (i) an orthogonal basis for S, and (ii) an orthonormal basis for S.

$6 \bullet$ Linear Transformations

6.1 Linear transformations

We begin by giving definitions concerning linear transformations in terms of $\mathbb{R}^n$, and proceed by applying these ideas first to $\mathbb{R}^2$ and then to $\mathbb{R}^3$. Again we shall use some ideas from linear algebra without always including proofs, but again these can be found in any good book on elementary linear algebra. We shall concentrate upon the geometrical aspects of linear transformations.

● Definition 1

A linear transformation $t : V \to W$ is a function from a vector space V to a vector space W which satisfies the following conditions for all $\mathbf{u}$ and $\mathbf{v}$ in V and all $\alpha \in \mathbb{R}$:

$$\text{(i)} \quad t(\mathbf{u} + \mathbf{v}) = t(\mathbf{u}) + t(\mathbf{v}) \quad \text{and} \quad \text{(ii)} \quad t(\alpha\mathbf{u}) = \alpha t(\mathbf{u}).$$

An equivalent definition combines these two requirements into a single more complicated requirement, as follows:

● Definition 2

A linear transformation $t : V \to W$ is a function from a vector space V to a vector space W which satisfies the following conditions for all $\mathbf{u}$ and $\mathbf{v}$ in V and all $\alpha, \beta \in \mathbb{R}$:

$$t(\alpha\mathbf{u} + \beta\mathbf{v}) = \alpha t(\mathbf{u}) + \beta t(\mathbf{v}).$$

● Definition 3

If in either of the above definitions $V = W = \mathbb{R}^n$ then we call the linear transformation a *linear transformation of* $\mathbb{R}^n$.

6.2 Linear transformations of $\mathbb{R}^2$

Let us start by looking at linear transformations of $\mathbb{R}^2$ since these are easy to visualise geometrically. Consider, for example, reflection of $\mathbb{R}^2$ in the x-axis.

We can see from Fig 6.1 that the point whose position vector is $\begin{pmatrix} x \\ y \end{pmatrix}$ is mapped onto the point with position vector $\begin{pmatrix} x \\ -y \end{pmatrix}$ and we could write this as $\begin{pmatrix} x \\ y \end{pmatrix}$ is mapped to the point $\begin{pmatrix} x' \\ y' \end{pmatrix}$, where

$$\begin{pmatrix} x' \\ y' \end{pmatrix} = \begin{pmatrix} 1 & 0 \\ 0 & -1 \end{pmatrix} \begin{pmatrix} x \\ y \end{pmatrix}.$$

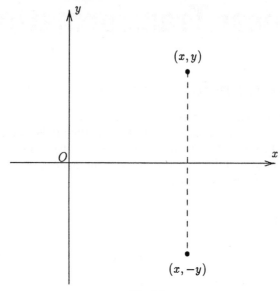

Fig 6.1

Does this mean that matrices will be of use here? Indeed it does, for if $A = \begin{pmatrix} a & b \\ c & d \end{pmatrix}$ is any 2×2 matrix with real entries, and if $t : \mathbb{R}^2 \to \mathbb{R}^2$ is defined by

$$t(\mathbf{x}) = A\mathbf{x} \quad \text{where} \quad \mathbf{x} = \begin{pmatrix} x \\ y \end{pmatrix},$$

then by matrix algebra

$$A(\mathbf{x} + \mathbf{w}) = A\mathbf{x} + A\mathbf{w} \quad \text{and} \quad A(\alpha\mathbf{x}) = \alpha A\mathbf{x}.$$

This means that any matrix transformation of this type is a linear transformation.

Now suppose we have a linear transformation which maps $\begin{pmatrix} 1 \\ 0 \end{pmatrix}$ to $\begin{pmatrix} a \\ c \end{pmatrix}$, and $\begin{pmatrix} 0 \\ 1 \end{pmatrix}$ to $\begin{pmatrix} b \\ d \end{pmatrix}$. Then since

$$\begin{pmatrix} x \\ y \end{pmatrix} = x \begin{pmatrix} 1 \\ 0 \end{pmatrix} + y \begin{pmatrix} 0 \\ 1 \end{pmatrix},$$

by using the properties of linear transformations,

$$t\left(\begin{pmatrix} x \\ y \end{pmatrix}\right) = xt\left(\begin{pmatrix} 1 \\ 0 \end{pmatrix}\right) + yt\left(\begin{pmatrix} 0 \\ 1 \end{pmatrix}\right)$$

$$= x \begin{pmatrix} a \\ c \end{pmatrix} + y \begin{pmatrix} b \\ d \end{pmatrix}$$

$$= \begin{pmatrix} a & b \\ c & d \end{pmatrix} \begin{pmatrix} x \\ y \end{pmatrix}.$$

This means that any linear transformation from $\mathbb{R}^2$ to $\mathbb{R}^2$ is a matrix transformation of the type described above. By analogous arguments, this leads to the following theorem.

● *Theorem I* ────────────────────────

A function $t : \mathbb{R}^n \to \mathbb{R}^m$ is a linear transformation if and only if it is a matrix transformation of the type

$$t(\mathbf{v}) = A\mathbf{v},$$

where A is a real $m \times n$ matrix.

● *Corollary I* ───────────────────────

A linear transformation $t : \mathbb{R}^n \to \mathbb{R}^m$ maps the zero vector of $\mathbb{R}^n$ to the zero vector of $\mathbb{R}^m$.

PROOF
Since from the theorem $t(\mathbf{v}) = A\mathbf{v}$, we know that $t(\mathbf{0}) = A\mathbf{0} = \mathbf{0}$.

From the working above, we see that the columns of the matrix representing a linear transformation of $\mathbb{R}^2$ are given by the images of $\begin{pmatrix} 1 \\ 0 \end{pmatrix}$ and $\begin{pmatrix} 0 \\ 1 \end{pmatrix}$ respectively. Likewise, in $\mathbb{R}^3$ the columns of the 3×3 matrix representing a linear transformation of $\mathbb{R}^3$ are the images of $\begin{pmatrix} 1 \\ 0 \\ 0 \end{pmatrix}$, $\begin{pmatrix} 0 \\ 1 \\ 0 \end{pmatrix}$ and $\begin{pmatrix} 0 \\ 0 \\ 1 \end{pmatrix}$ respectively.

6.3 Some special linear transformations of $\mathbb{R}^2$

Rotation about *O*

Under rotation through an angle θ about the origin (measuring θ anticlockwise from the positive x-axis), the images of the points $(1,0)$ and $(0,1)$ will be $(\cos\theta, \sin\theta)$ and $(-\sin\theta, \cos\theta)$ repectively as shown in Fig 6.2. The respective columns of the matrix representing the rotation are given by the position vectors of these points, and hence the matrix is

$$\begin{pmatrix} \cos\theta & -\sin\theta \\ \sin\theta & \cos\theta \end{pmatrix}.$$

Reflection in a Line through *O*

We looked at a reflection in the x-axis, but suppose we have reflection in a line making an angle θ with the x-axis.

From Fig 6.3, we see that the point $(1,0)$ is mapped onto the point $(\cos 2\theta, \sin 2\theta)$, and the point $(0,1)$ is mapped onto the point

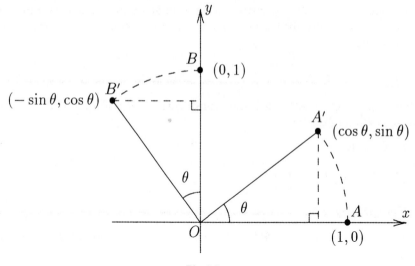

Fig 6.2

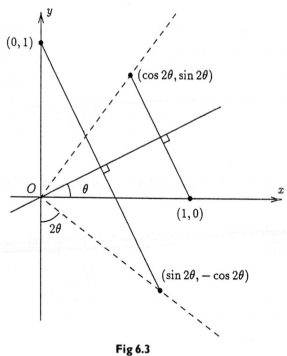

Fig 6.3

$(\sin(\pi - 2\theta), \cos(\pi - 2\theta))$, which can be written as $(\sin 2\theta, -\cos 2\theta)$, so the matrix representing reflection in a line making an angle θ with the positive x-axis is

$$\begin{pmatrix} \cos 2\theta & \sin 2\theta \\ \sin 2\theta & -\cos 2\theta \end{pmatrix}.$$

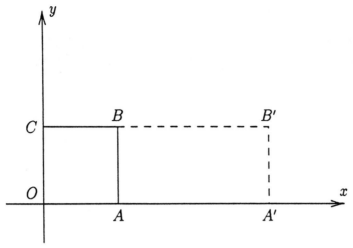

Fig 6.4

Stretch Parallel to the Axes

Another type of familiar transformation of $\mathbb{R}^2$ is the *stretch parallel to the x-axis*. In this case all distances from the y-axis are multiplied by a constant factor α, say, where, if α is positive, the image is the same side of the y-axis as the original point, whilst if α is negative, the image point is on the opposite side of the y-axis to the original point. We call α the *scale factor* of the stretch, and if it is a stretch in this direction only the matrix will be

$$\begin{pmatrix} \alpha & 0 \\ 0 & 1 \end{pmatrix},$$

but if there is also a stretch parallel to the y-axis, with scale factor β, our matrix will be

$$\begin{pmatrix} \alpha & 0 \\ 0 & \beta \end{pmatrix}.$$

If $|\alpha| < 1$ then image points will be closer to the y-axis than the original points, and if $|\alpha| > 1$ then image points will be farther away from the y-axis than the original points. A similar rule will apply for β.

Enlargement

A special case of this last transformation occurs when $\alpha = \beta$. This will be an enlargement, and since the scale factor is the same in both x- and y-directions, the scaling in any direction will be the same value. Here the matrix will be of the form

$$\begin{pmatrix} \alpha & 0 \\ 0 & \alpha \end{pmatrix}.$$

Shear

A shear is a transformation which keeps all points on one line through the origin fixed, but moves all other points parallel to the fixed line. The distance each point moves is proportional to its distance from the fixed line. We look at the special case where this fixed line is the x-axis. Then the matrix is of the form

$$\begin{pmatrix} 1 & k \\ 0 & 1 \end{pmatrix}, \quad k \neq 0$$

and it represents the transformation which sends any point (x, y) to the point $(x + ky, y)$, so that the distance moved is proportional to the y-coordinate. In Fig 6.5 $k = 2$.

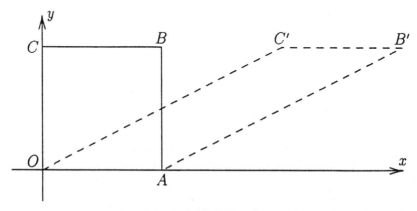

Fig 6.5

Identity Transformation

If in the case of a shear we allow k to be zero, then we get the identity matrix, which represents the identity transformation. This leaves every point of $\mathbb{R}^2$ fixed and is also a special case of an enlargement (with $\alpha = 1$) and of a rotation (where the angle turned through is any multiple of 2π).

Singular Transformations

In all of the above transformations, a non-zero volume is mapped onto a non-zero volume, but this need not necessarily be the case. Consider the linear transformation t which projects the whole of $\mathbb{R}^2$ onto a line in $\mathbb{R}^2$. For example, if t has matrix

$$\begin{pmatrix} pa & pb \\ qa & qb \end{pmatrix}$$

then

$$\begin{pmatrix} pa & pb \\ qa & qb \end{pmatrix} \begin{pmatrix} x \\ y \end{pmatrix} = \begin{pmatrix} pax + pby \\ qax + qby \end{pmatrix} = (ax + by) \begin{pmatrix} p \\ q \end{pmatrix}.$$

In this case every point in $\mathbb{R}^2$ is mapped to a point on the line $py = qx$, and thus the whole plane *collapses* down onto this line. There is one exceptional case amongst these, when $p = q = 0$ (or $a = b = 0$), and in this case the whole of the plane collapses down onto the single point at the origin. A transformation of this type is called a *singular transformation* because its matrix is *singular* (that is, it has zero determinant).

Why have we looked at these special cases? It is because any linear transformation can be thought of as a combination of these special cases.

EXERCISE 1

(i) Write down the matrices for the following transformations where all angles will be given in radians. (Recall that π radians $= 180°$.)

(a) rotation about O through $\pi/3$,
(b) reflection in the line through O making an angle $\pi/6$ with the x-axis,
(c) a shear, keeping the y-axis fixed, and sending the point $(1,0)$ to the point $(1,3)$.

(ii) For each of the matrices found in (i) calculate the images of the points $(1,0)$ and $(0,1)$.

(iii) For each transformation in (i) sketch the points $(1,0)$ and $(0,1)$, their images under the action of the corresponding linear transformation and check whether your diagrams agree with your answers to (ii).

6.4 Combinations of linear transformations

Suppose t and s are two linear transformations of $\mathbb{R}^2$ with respective matrices A and B. Let $\mathbf{x}'$ be the image of $\mathbf{x}$ under t and $\mathbf{x}''$ be the image of $\mathbf{x}'$ under s. Then if we perform first t and then s, $\mathbf{x}$ will be mapped to $\mathbf{x}''$. Since

$$\mathbf{x}' = t(\mathbf{x}) = A\mathbf{x} \quad \text{and} \quad \mathbf{x}'' = s(\mathbf{x}') = B\mathbf{x}'$$

we have

$$\mathbf{x}'' = B(A\mathbf{x}) = BA\mathbf{x}.$$

This shows that the combination transformation t followed by s, which we shall refer to as st, is a linear transformation with matrix BA. Notice the order: t acts first so is written on the right so that it can *act on* the vector $\mathbf{x}$ first, after which s acts on the resulting vector $\mathbf{x}'$. Another way of expressing this is

$$st(\mathbf{x}) = s(t(\mathbf{x})).$$

EXERCISE 2

Consider the following linear transformations:

(i) s is a rotation about the origin through $\pi/3$,
(ii) t is a rotation about the origin through $\pi/6$,
(iii) p is a reflection in a line though the origin making an angle $\pi/6$ with the x-axis,
(iv) q is a reflection in a line though the origin making an angle $\pi/3$ with the x-axis.

Find the matrices for each of these transformations, and the matrix for

(a) st, (b) sp, (c) ps and (d) pq.

This exercise illustrates the facts that

(i) Successive rotations through angles θ and ϕ about O result in a rotation about O through an angle $\theta + \phi$.

(ii) A reflection in a line through O at an angle ϕ followed by a rotation about O through an angle θ results in a reflection in a line through O at an angle $\phi + \theta/2$.

(iii) A rotation about O through an angle θ followed by a reflection in a line through O at an angle ϕ results in a reflection in a line through O at an angle $\phi - \theta/2$.

(iv) A reflection in a line through O at an angle ϕ followed by a reflection in a line through O at an angle θ results in a rotation about O through an angle $2(\theta - \phi)$.

These general results can be verified by multiplying out the respective matrices of reflections and rotations in the given combinations. A proof for result (ii) is given in Example 1 below, but proofs for (i), (iii) and (iv) are given as Exercise 10 in the end-of-chapter exercises.

◉ *Example I*

The matrices for a reflection in a line through O at an angle ϕ followed by a rotation about O through an angle θ are respectively

$$\begin{pmatrix} \cos 2\phi & \sin 2\phi \\ \sin 2\phi & -\cos 2\phi \end{pmatrix} \quad \text{and} \quad \begin{pmatrix} \cos \theta & -\sin \theta \\ \sin \theta & \cos \theta \end{pmatrix}.$$

Hence the matrix for the reflection followed by the rotation is

$$\begin{pmatrix} \cos \theta & -\sin \theta \\ \sin \theta & \cos \theta \end{pmatrix} \begin{pmatrix} \cos 2\phi & \sin 2\phi \\ \sin 2\phi & -\cos 2\phi \end{pmatrix}$$
$$= \begin{pmatrix} \cos \theta \cos 2\phi - \sin \theta \sin 2\phi & \cos \theta \sin 2\phi + \sin \theta \cos 2\phi \\ \sin \theta \cos 2\phi + \cos \theta \sin 2\phi & \sin \theta \sin 2\phi - \cos \theta \cos 2\phi \end{pmatrix}$$
$$= \begin{pmatrix} \cos(2\phi + \theta) & \sin(2\phi + \theta) \\ \sin(2\phi + \theta) & -\cos(2\phi + \theta) \end{pmatrix}.$$

This is the matrix for reflection in a line through O at an angle $(\phi + \theta/2)$ which proves our result.

6.5 Fixed lines, eigenvectors and eigenvalues

From Corollary 1, we know that the origin is *fixed* (that is, mapped to itself) under any linear transformation. If we consider the special cases of the previous section, then in most cases there are lines which are *fixed*.

● *Definition 4*

A line l is *fixed* under a linear transformation t of $\mathbb{R}^n$ if t maps any point X on l to a point X' also on l. (X and X' are not necessarily the same point, although this is a possibility.)

● Definition 5

Let t be a linear transformation of $\mathbb{R}^n$ with corresponding matrix A. Then, if

$$A\mathbf{v} = \lambda\mathbf{v}$$

$\mathbf{v}$ is called an *eigenvector* of t with corresponding *eigenvalue* λ of t. Sometimes these are referred to as eigenvector and corresponding eigenvalue of the matrix A.

● Theorem 2 ————————————————

If $\mathbf{v}$ is an eigenvector of a linear transformation t of $\mathbb{R}^n$, with corresponding eigenvalue λ, then for any non-zero real number k, $k\mathbf{v}$ is an eigenvector of t with corresponding eigenvalue λ.

PROOF
Let A be the matrix of t; then $A\mathbf{v} = \lambda\mathbf{v}$ by the assumption, so that

$$t(k\mathbf{v}) = A(k\mathbf{v}) = k(A\mathbf{v}) = k(\lambda\mathbf{v}) = \lambda(k\mathbf{v})$$

and the theorem is proved.

● Corollary 2 ————————————————

If $\mathbf{v}$ is an eigenvector, with corresponding eigenvalue λ, of a linear transformation t of $\mathbb{R}^n$, then the line l which passes through the origin and the point V, with position vector $\mathbf{v}$, is a fixed line of t and the position vector of any point on l, other than 0, is an eigenvector of t with eigenvalue λ.

PROOF
Any point on l must have position vector $k\mathbf{v}$ for some $k \in \mathbb{R}$, and the result follows immediately from the theorem.

This means that on the fixed line l, λ is the scale factor by which all distances from the origin are multiplied under the action of t.

● Example 2

Suppose t is a linear transformation of $\mathbb{R}^2$ with matrix $\begin{pmatrix} 1 & 2 \\ 2 & 1 \end{pmatrix}$. Show that $\begin{pmatrix} 1 \\ 1 \end{pmatrix}$ and $\begin{pmatrix} 1 \\ -1 \end{pmatrix}$ are eigenvectors of t, and find their corresponding eigenvalues. Verify that $\begin{pmatrix} k \\ k \end{pmatrix}$ is also an eigenvector of t with the same eigenvalue as $\begin{pmatrix} 1 \\ 1 \end{pmatrix}$ and hence that $y = x$ is a fixed line of t. What is the equation of another fixed line of t?

SOLUTION

$$\begin{pmatrix} 1 & 2 \\ 2 & 1 \end{pmatrix}\begin{pmatrix} 1 \\ 1 \end{pmatrix} = \begin{pmatrix} 3 \\ 3 \end{pmatrix} = 3\begin{pmatrix} 1 \\ 1 \end{pmatrix}.$$

Hence from the definition, $\begin{pmatrix} 1 \\ 1 \end{pmatrix}$ is an eigenvector with corresponding eigenvalue 3. Similarly

$$\begin{pmatrix} 1 & 2 \\ 2 & 1 \end{pmatrix} \begin{pmatrix} 1 \\ -1 \end{pmatrix} = \begin{pmatrix} -1 \\ 1 \end{pmatrix} = (-1) \begin{pmatrix} 1 \\ -1 \end{pmatrix},$$

so $\begin{pmatrix} 1 \\ -1 \end{pmatrix}$ is an eigenvector with corresponding eigenvalue -1. Similarly

$$\begin{pmatrix} 1 & 2 \\ 2 & 1 \end{pmatrix} \begin{pmatrix} k \\ k \end{pmatrix} = \begin{pmatrix} 3k \\ 3k \end{pmatrix} = 3 \begin{pmatrix} k \\ k \end{pmatrix}.$$

Again $\begin{pmatrix} k \\ k \end{pmatrix}$ is an eigenvector with eigenvalue 3 for any $k \in \mathbb{R}$.

Now a point lies on $y = x$ if and only if its position vector is of the form $\begin{pmatrix} k \\ k \end{pmatrix}$ and any such vector is mapped to the position vector of another point on $y = x$, so $y = x$ is a fixed line of t. Since a similar result holds for $\begin{pmatrix} k \\ -k \end{pmatrix}$ another fixed line of t is $y = -x$.

Consider the linear transformation t whose matrix is $\begin{pmatrix} a & b \\ c & d \end{pmatrix}$. Then if l is a fixed line of t passing through the origin, and (x, y) is a point on l, for some $\lambda \in r$ we have

$$\lambda \begin{pmatrix} x \\ y \end{pmatrix} = \begin{pmatrix} a & b \\ c & d \end{pmatrix} \begin{pmatrix} x \\ y \end{pmatrix}$$

or

$$\lambda x = ax + by \tag{6.5.1}$$
$$\lambda y = cx + dy. \tag{6.5.2}$$

By eliminating x and y from these equations we get

$$\lambda^2 - (a + d)\lambda + (ad - bc) = 0. \tag{6.5.3}$$

From the theory of quadratic equations, we know that if α and β are the roots of the quadratic equation

$$px^2 + qx + r = 0$$

then the product of the roots is equal to r/p, and the sum of the roots is equal to $-q/p$. Thus in equation (6.5.3) we see that the product of the roots is $ad - bc$, which means that the product of the eigenvalues of a linear transformation t of $\mathbb{R}^2$ is equal to the determinant of the matrix of t. Because, when the eigenvalues are distinct and real, they give the scale factors of t in two given directions, their product gives the scale factor of the effect of t on any area. We shall see that, even if the eigenvalues are not real, or not distinct, the determinant of the matrix of t still gives the scale factor of the effect of t on any area.

• *Example 3*

(i) Find the eigenvalues and eigenvectors of the transformation of $\mathbb{R}^2$ whose matrix is $\begin{pmatrix} 3 & 2 \\ 2 & 3 \end{pmatrix}$, and hence illustrate the transformation by means of a diagram.

(ii) Repeat the process for the transformation whose matrix is $\begin{pmatrix} 3 & 7 \\ 2 & -2 \end{pmatrix}$.

SOLUTION

(i) By using equation (6.5.3) we find that the eigenvalues of the transformation are the solutions of the equation

$$\lambda^2 - 6\lambda + 5 = 0.$$

The solutions are $\lambda = 1$ and $\lambda = 5$, and by substituting these values into equations (6.5.1) and (6.5.2) we find that the equation of the fixed line corresponding to $\lambda = 1$ is $y = -x$, and that corresponding to $\lambda = 5$ is $y = x$. Thus our transformation keeps distances in the direction of the line $y = -x$ fixed, and stretches distances by a factor of 5 in the direction of $y = x$, as shown in Fig 6.6.

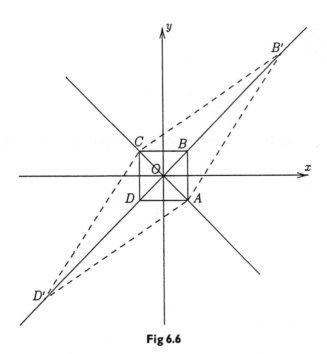

Fig 6.6

(ii) In the second case we find that the solutions to equation (6.5.3) are $\lambda = -4$ and $\lambda = 5$, and the corresponding fixed lines are respectively $y = -x$ and $7y = 2x$. Thus distances in the direction of $y = -x$ are multiplied by -4, which means there is a reflection of the line in the origin as well as a scaling by a factor of 4, whilst in the direction of $7y = 2x$ there is simply a scaling by a factor of 5, as shown in Fig 6.7.

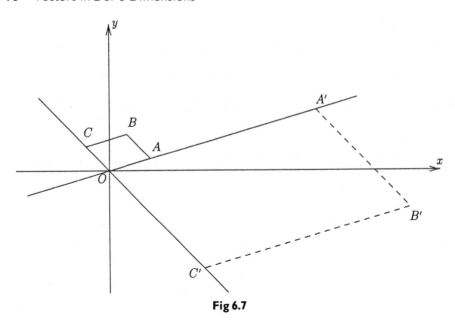

Fig 6.7

EXERCISE 3

Find the eigenvalues and eigenvectors of the transformation of $\mathbb{R}^2$ whose matrix is $\begin{pmatrix} 2 & 2 \\ 2 & -1 \end{pmatrix}$, and hence illustrate the transformation by means of a diagram.

6.6 Eigenvectors and eigenvalues in special cases

Rotation about *O*

In this case $a = d = \cos\theta$ and $-b = c = \sin\theta$, so that equation (6.5.3) becomes

$$\lambda^2 - 2\cos\theta\lambda + \cos^2\theta + \sin^2\theta = 0$$

which can be written as

$$(\lambda - \cos\theta)^2 + \sin^2\theta = 0.$$

If $\sin\theta \neq 0$ this has no real solutions, and the rotation has no fixed lines, but if $\sin\theta = 0$, then $\lambda = \cos\theta = \pm 1$ and our rotation is through a multiple of π. If θ is an even multiple of π then $\lambda = 1$, but if θ is an odd multiple of π then $\lambda = -1$. In both cases the determinant of the matrix is 1, as we should expect, since a rotation leaves areas unchanged.

EXERCISE 4

Draw a diagram to illustrate the points (1,0) and (0,1) in $\mathbb{R}^2$ and their images under rotation through (i) π, and (ii) 2π, and explain why the corresponding eigenvalues would be -1 and 1 respectively.

Reflection in a Line through **O**

Suppose the line makes an angle θ with the x-axis (measured anticlockwise). In this case $a = -d = \cos 2\theta$ and $b = c = \sin 2\theta$. In this case equation (6.5.3) becomes

$$\lambda^2 - \cos^2 2\theta - \sin^2 2\theta = 0$$

which reduces to

$$\lambda^2 - 1 = 0.$$

So λ can take either the value 1 or -1.

If $\lambda = 1$, from equations (6.5.1) and (6.5.2), we get

$$x = (\cos 2\theta)x + (\sin 2\theta)y.$$

By considering the formulae for double angles

$$\sin 2\theta = 2\sin\theta\cos\theta \quad \text{and} \quad \cos 2\theta = 1 - 2\sin^2\theta$$

and by cancelling $2\sin\theta$ we get

$$(\cos\theta)y = (\sin\theta)x,$$

which is the equation of the line making an angle θ with the positive x-axis. This should come as no great surprise, as, under reflection in l, we should expect all points on l to stay exactly where they are, and the fact that $\lambda = 1$ on this line ensures that this is the case.

If $\lambda = -1$, equations (6.5.1) and (6.5.2) give

$$-x = (\cos 2\theta)x + (\sin 2\theta)y,$$

and by using the identity $\cos 2\theta = 2\cos^2\theta - 1$ and the identity for $\sin 2\theta$ we arrive at

$$(-\sin\theta)y = (\cos\theta)x,$$

which is the equation of the line though the origin making an angle $\theta + \pi/2$ with the x-axis, and therefore perpendicular to l. Again this should not be surprising since, under reflection in l, any point P on the line l' at right angles to l is mapped onto a point P' on l' on the opposite side of O from P, where the distances OP and OP' are equal. Because of the flip over to the other side the coordinates are multiplied by -1, the scale factor in this case.

For a reflection of $\mathbb{R}^2$ in a line through the origin, areas are unchanged in *size*, but their orientation is *reversed*. That is, if a triangle has its vertices A, B, C labelled anticlockwise, under the reflection the corresponding images A', B', C' will go round the image triangle in a clockwise direction. It is true to say that a negative determinant of the matrix of a linear transformation indicates that orientation is reversed, whilst a positive determinant indicates that orientation is preserved.

EXERCISE 5

Write down the matrix for reflection in the line $y = x$. (What angle does this make with the positive x-axis?) What is the equation of the line through O at right angles to $y = x$? Choose a point other than the origin on each line and find its image under this transformation by using the matrix. Then sketch the points you chose, their images under the given reflection and the two lines mentioned above, and check that your geometric and algebraic results agree with one another.

Stretches Parallel to the Coordinate Axes

In any of the cases where $b = c = 0$, with $a = \alpha$, $d = \beta$ with $\alpha \neq \beta$, our equation (6.5.3) becomes

$$\lambda^2 - (\alpha + \beta)\lambda + \alpha\beta = 0$$

giving $\lambda = \alpha$ or $\lambda = \beta$, and the fixed lines are the axes of coordinates with scale factor α in the x-direction and scale factor β in the y-direction – as expected.

Enlargements

In the previous case where $\alpha = \beta$ we get $\lambda = \alpha$ as the only solution and equations (6.5.1) and (6.5.2) are true for all values of x and y. This means that every line through the origin is a fixed line and the scale factor along each is α. If a figure has any of its linear measurements multiplied by a factor of α then its area is multiplied by α^2, the value of the determinant.

Shears

We next look at shears where $a = d = 1$, $c = 0$ and $b = k$. Equation (6.5.3) becomes

$$\lambda^2 - 2\lambda + 1 = 0$$

which means that $\lambda = 1$, and equations (6.5.1) and (6.5.2) tell us that $y = 0$, so that the only fixed line through the origin is the x-axis, and every point on the x-axis is a fixed point. (In fact, in this case, all lines parallel to the x-axis will be fixed lines, and all points on one of these fixed lines will undergo a translation parallel to the x-axis, but the magnitude of that translation will be proportional to the distance of the line from the x-axis. Note that eigenvectors only tell us which are the fixed lines *through the origin*.) Notice that under a shear, shapes may change, but areas are preserved, which is what we should expect with the matrix having a determinant of 1.

Singular Transformations

The matrices of these transformations are singular matrices. That is, they have zero determinant. If the matrix is $\begin{pmatrix} pa & pb \\ qa & qb \end{pmatrix}$, then equation (6.5.3) becomes

$$\lambda^2 - (pa + qb)\lambda = 0$$

since $paqb - qapb = 0$. If $pa + qb \neq 0$, the solutions are

$$\lambda = 0 \qquad \text{and} \qquad \lambda = pa + qb$$

and by substituting into equations (6.5.1) and (6.5.2) we find that the eigenvalue $\lambda = 0$ corresponds to an eigenvector in the direction of the line $ax + by = 0$, and eigenvalue $\lambda = pa + qb$ corresponds to the line $py = qx$. Note that these lines are not necessarily perpendicular.

In the case where $pa + qb = 0$, all points are mapped onto the origin.

6.7 **Linear transformations of** $\mathbb{R}^3$

Eigenvalues and eigenvectors were defined in Section 6.5 for any $\mathbb{R}^n$, so results in $\mathbb{R}^3$ are analogous to those in $\mathbb{R}^2$. Rather than plod through a mass of theory, a brief explanation of how eigenvalues and eigenvectors are normally found will be given here, but we shall go on to examine the three-dimensional cases by means of particular examples chosen to illustrate geometrical points.

From Definition 5 we know that if $\mathbf{v}$ is an eigenvector of t whose matrix is A, and if λ is the corresponding eigenvalue, then

$$A\mathbf{v} = \lambda\mathbf{v}.$$

This could be written as

$$(A - \lambda I)\mathbf{v} = \mathbf{0}, \tag{6.7.1}$$

and linear algebra theory tells us that this only has a non-zero vector solution if the determinant of $(A - \lambda I)$ is zero, that is if

$$|A - \lambda I| = 0. \tag{6.7.2}$$

The reason for this is that, if the determinant were non-zero, then the matrix $(A - \lambda I)$ would have an inverse, which in turn would mean that if we multiply both sides of equation (6.7.1) on the left by this inverse we should get

$$\mathbf{v} = (A - \lambda I)^{-1}\mathbf{0} = \mathbf{0}.$$

But $\mathbf{v}$ cannot be the zero vector, since by definition an eigenvector is a *non-zero* vector.

● *Definition 6*

Equation (6.7.2) is called the *characteristic equation* of the matrix A.

By solving equation (6.7.2) for λ, we can substitute each solution for λ back into the matrix equation (6.7.1) to find a corresponding eigenvector, and these eigenvectors will give the directions of the fixed lines through the origin.

In $\mathbb{R}^2$ it is easy to show that equation (6.7.2) is equivalent to equation (6.5.3) as follows. If $A = \begin{pmatrix} a & b \\ c & d \end{pmatrix}$, then

$$|A - \lambda I| = \left| \begin{pmatrix} a - \lambda & b \\ c & d - \lambda \end{pmatrix} \right|$$

so that equation (6.7.1) becomes

$$(a - \lambda)(d - \lambda) - bc = 0,$$

which, when multiplied out becomes

$$\lambda^2 - (a + d)\lambda + ad - bc = 0$$

as predicted.

It is still true that the product of the eigenvalues is equal to the determinant of the matrix of the linear transformation, but now this value represents the scale factor by which volume is multiplied under the linear transformation of $\mathbb{R}^3$.

EXERCISE 6

Find the characteristic equation for each of the following matrices.

(a) $\begin{pmatrix} 1 & 2 \\ 3 & 4 \end{pmatrix}$ (b) $\begin{pmatrix} 3 & 4 \\ 4 & -3 \end{pmatrix}$ (c) $\begin{pmatrix} 3 & -4 \\ 4 & 3 \end{pmatrix}$ (d) $\begin{pmatrix} 1 & 2 \\ 4 & 8 \end{pmatrix}$.

6.8 Special cases in $\mathbb{R}^3$

Rotation about an Axis through the Origin

In this case the axis is a line of fixed points, so 1 is an eigenvalue with any vector along this axis being a corresponding eigenvector. The other eigenvalues will be complex, and so there will be no other eigenvector directions. However, since a rotation keeps volume and orientation fixed, the 3×3 matrix of the rotation will have determinant 1.

Reflection in a Plane through the Origin

The eigenvalues for a reflection in a plane through the origin in $\mathbb{R}^3$ will be 1, 1 and -1, since there will be a whole plane of fixed points, and any vector along the line through the origin perpendicular to this fixed plane will be reflected to a vector of the same length but in the opposite direction to the original vector. The eigenvalue 1 will correspond to all the vectors in the plane of reflection, and the eigenvalue -1 corresponds to vectors along the line through the origin, perpendicular to this plane of reflection. The determinant of the matrix corresponding to this reflection is -1, and equivalently the product of the eigenvalues is -1.

Stretches, Enlargements and Shears

Again these will be analogous to the two-dimensional cases. If the linear transformation t has eigenvalues α, β, γ corresponding to eigenvectors $\mathbf{u}, \mathbf{v}, \mathbf{w}$ respectively, then t is equivalent to stretches by scale factors α, β, γ in the directions (not necessarily orthogonal) of $\mathbf{u}, \mathbf{v}, \mathbf{w}$. In each case, the product of the eigenvalues will be equal to the determinant of the matrix of t. Enlargements will involve three equal eigenvalues, and any vector in $\mathbb{R}^3$ will be an eigenvector of an enlargement centred at the origin. There may be shears in one or two directions, the latter transforming a rectangular block into a parallelepiped.

Example 4

Suppose t is the linear transformation of $\mathbb{R}^3$ with matrix

$$A = \begin{pmatrix} 4 & 1 & 1 \\ 1 & 4 & 1 \\ 1 & 1 & 4 \end{pmatrix}.$$

Show that the line $x = y = z$ is a fixed line of t, and that the scale factor of the transformation on this line is 6. Show also that the transformation acts as an enlargement with scale factor 3 on the plane $x + y + z = 0$.

SOLUTION

Any point on the line $x = y = z$ has position vector of the form $\begin{pmatrix} k \\ k \\ k \end{pmatrix}$. Thus

$$\begin{pmatrix} 4 & 1 & 1 \\ 1 & 4 & 1 \\ 1 & 1 & 4 \end{pmatrix} \begin{pmatrix} k \\ k \\ k \end{pmatrix} = \begin{pmatrix} 4k + k + k \\ k + 4k + k \\ k + k + 4k \end{pmatrix} = 6 \begin{pmatrix} k \\ k \\ k \end{pmatrix}.$$

Hence $\begin{pmatrix} k \\ k \\ k \end{pmatrix}$ is an eigenvector of t with corresponding eigenvalue 6. Thus $x = y = z$ is a fixed line along which the scale factor of t is 6.

Now consider the plane $x + y + z = 0$. Since in this case $z = -(x + y)$, any point

on this plane has position vector $\begin{pmatrix} x \\ y \\ -(x+y) \end{pmatrix}$, and

$$\begin{pmatrix} 4 & 1 & 1 \\ 1 & 4 & 1 \\ 1 & 1 & 4 \end{pmatrix} \begin{pmatrix} x \\ y \\ -(x+y) \end{pmatrix} = \begin{pmatrix} 4x + y - (x+y) \\ x + 4y - (x+y) \\ x + y - 4(x+y) \end{pmatrix} = 3 \begin{pmatrix} x \\ y \\ -(x+y) \end{pmatrix}$$

which means that every vector in the plane is an eigenvector with corresponding eigenvalue 3, and hence t acts on the plane as an enlargement with scale factor 3.

EXERCISE 7

Suppose t is the linear transformation of $\mathbb{R}^3$ whose matrix is

$$A = \begin{pmatrix} 1 & 1 & 4 \\ 1 & 4 & 1 \\ 4 & 1 & 1 \end{pmatrix}.$$

(i) Show that $\begin{pmatrix} 1 \\ -2 \\ 1 \end{pmatrix}$, $\begin{pmatrix} 1 \\ 1 \\ 1 \end{pmatrix}$ and $\begin{pmatrix} 1 \\ 0 \\ -1 \end{pmatrix}$ are eigenvectors for t, and find their

corresponding eigenvalues.

(ii) Solve equation (6.7.2) for the A given above, and for each of the solutions for λ

find corresponding eigenvectors by solving the equation (6.7.1) where $\mathbf{x} = \begin{pmatrix} x \\ y \\ z \end{pmatrix}$.

Check your answer to (ii) with that from (i).

Summary

1. A linear transformation of $\mathbb{R}^n$ was described in Section 6.1.
2. A function $t : \mathbb{R}^n \to \mathbb{R}^m$ is a linear transformation if and only if it is a matrix transformation of the form

$$t(\mathbf{v}) = A\mathbf{v},$$

where A is a real $m \times n$ matrix.

3. Examples of linear transformations of $\mathbb{R}^2$ (or $\mathbb{R}^3$) are rotations about the origin, reflections in lines (or planes) through the origin, stretches parallel to axes, enlargements, shears, projections or combinations of these. The *identity transformation* leaves every point fixed.
4. Suppose t is a linear transformation of $\mathbb{R}^n$ with corresponding matrix A, and suppose also that

$$A\mathbf{v} = \lambda\mathbf{v};$$

then $\mathbf{v}$ is an *eigenvector* of t, and λ is its corresponding eigenvalue.
5. The *characteristic equation* of a square matrix A is

$$|A - \lambda I| = 0$$

and the solutions to this equation are the eigenvalues of A.
6. Eigenvectors of a linear transformation determine the directions of the fixed lines through the origin under that transformation. The corresponding eigenvalues give the scale factor of the transformation along these fixed lines.
7. The product of the eigenvalues of a linear transformation t of $\mathbb{R}^n$ is equal to the determinant of the matrix of t.

FURTHER EXERCISES

8. Find the matrices of the linear transformations p, q, r, s, t, u of $\mathbb{R}^2$ where

    p is an enlargement centred at the origin with scale factor 2,
    q is a rotation about the origin through an angle $\pi/2$,
    r is a reflection in the y-axis.
    s is a reflection in a line through the origin making an angle $\pi/4$ with the x-axis,
    t is a rotation about the origin through an angle $\pi/4$,
    u is a shear keeping the x-axis fixed and sending the point $(0,1)$ to the point $(3,1)$.

    In each case draw a diagram to illustrate both the unit square (whose vertices are $(0,0)$, $(1,0)$, $(1,1)$, $(0,1)$) and its image under the given transformation.
9. Using the transformations given in Exercise 8, find the matrices of the following transformations, and describe them geometrically:

$$qr, \quad rs, \quad sr, \quad st, \quad tr.$$

    Does this agree with the cases given in Section 6.4 concerning combinations of reflections and rotations?

10. By using a method similar to that in Example 1, or otherwise, prove that:
    (i) Successive rotations through angles θ and ϕ about O result in a rotation about O through an angle $\theta + \phi$.
    (ii) A rotation about O through an angle θ followed by a reflection in a line through O at an angle ϕ results in a reflection in a line through O at an angle $\phi - \theta/2$.
    (iii) A reflection in a line through O at an angle ϕ followed by a reflection in a line through O at an angle θ results in a rotation about O through an angle $2(\theta - \phi)$.

11. Find the characteristic equations, and hence the eigenvalues and corresponding fixed lines, of the linear transformations of $\mathbb{R}^2$ whose matrices are:

(i) $\begin{pmatrix} 1 & 2 \\ 1 & 0 \end{pmatrix}$,   (ii) $\begin{pmatrix} 3/5 & -4/5 \\ 4/5 & 3/5 \end{pmatrix}$,   (iii) $\begin{pmatrix} 1 & 0 \\ 2 & 1 \end{pmatrix}$,

(iv) $\begin{pmatrix} 3/5 & 4/5 \\ 4/5 & -3/5 \end{pmatrix}$,   (v) $\begin{pmatrix} 1 & 3 \\ 2 & 6 \end{pmatrix}$.

Describe each of these transformations geometrically.

12. Find the characteristic equation and hence the eigenvalues and fixed lines for each of the linear transformations of $\mathbb{R}^3$ whose matrices are:

(i) $\begin{pmatrix} 1 & -1 & 0 \\ 1 & 1 & 0 \\ 0 & 0 & 1 \end{pmatrix}$,   (ii) $\begin{pmatrix} 2 & 1 & -2 \\ 1 & -1 & 1 \\ -2 & 1 & 2 \end{pmatrix}$.

7 • General Reflections, Rotations and Translations in $\mathbb{R}^3$

In this chapter we consider transformations which do not necessarily keep the origin fixed and so will not be *linear* transformations. However, they are useful geometrically, and give excellent practice in the use of scalar products and vector products.

7.1 Reflections

We consider first reflection in a plane in $\mathbb{R}^3$. Suppose the point X whose position vector is $\mathbf{x}$ is reflected in the plane π whose equation is

$$(\mathbf{r} - \mathbf{a}).\mathbf{n} = 0$$

where $\mathbf{a}$ is the position vector of a point A in the plane, and $\mathbf{n}$ is a vector normal to the plane, and we shall choose $\mathbf{n}$ to be a unit vector for convenience. Let X' with position vector $\mathbf{x}'$ be the image of X under this reflection. Then the line segment

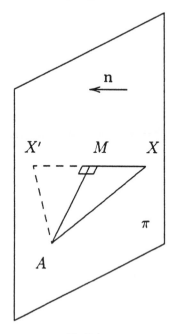

Fig 7.1

XX' is perpendicular to π, and if this line segment cuts π at M we have $\mathbf{XM} = ((\mathbf{a} - \mathbf{x}) . \mathbf{n})\mathbf{n}$, since $\mathbf{n}$ is a unit vector. Hence, as $\mathbf{XX'} = 2\mathbf{XM}$,

$$\mathbf{x}' - \mathbf{x} = 2((\mathbf{a} - \mathbf{x}) . \mathbf{n})\mathbf{n}$$

and

$$\mathbf{x}' = \mathbf{x} - 2((\mathbf{x} - \mathbf{a}) . \mathbf{n})\mathbf{n}. \tag{7.1.1}$$

Now if the whole of $\mathbb{R}^3$ is reflected in the plane, we can find the image of any point in $\mathbb{R}^3$ by using equation (7.1.1).

NOTE
If X lies on the plane π, then $\mathbf{x} - \mathbf{a}$ will be parallel to the plane and therefore perpendicular to $\mathbf{n}$. This means that $(\mathbf{x} - \mathbf{a}).\mathbf{n} = 0$, and hence that $\mathbf{x}' = \mathbf{x}$, as, indeed, we should expect, since any point on the plane of reflection is invariant under that reflection.

Example I

Find the image of the point $(2,1,3)$ under reflection in the plane whose equation is $2x - 2y + z = 1$.

SOLUTION
First we need a point on the plane, and $(0,0,1)$ satisfies the equation. Next a vector normal to the plane is $2\mathbf{i} - 2\mathbf{j} + \mathbf{k}$, and we need to divide this by its length to obtain a unit vector in this direction. So,

$$\mathbf{x} = 2\mathbf{i} + \mathbf{j} + 3\mathbf{k}, \quad \mathbf{a} = \mathbf{k}, \quad \mathbf{n} = \frac{2}{3}\mathbf{i} - \frac{2}{3}\mathbf{j} + \frac{1}{3}\mathbf{k}.$$

Now by direct substitution into equation (7.1.1) above we find that if X' is the image of X under this reflection

$$\mathbf{x}' = \frac{2}{9}\mathbf{i} + \frac{25}{9}\mathbf{j} + \frac{19}{9}\mathbf{k}.$$

Hence the image of the given point is $(2/9, 25/9, 19/9)$.

We can check this by showing that $\mathbf{x}' - \mathbf{x}$ is parallel to $\mathbf{n}$, that is perpendicular to the plane, and also by finding the midpoint of XX', and showing that it lies on the plane.

$$\mathbf{x}' - \mathbf{x} = \frac{8}{9}(2\mathbf{i} - 2\mathbf{j} + \mathbf{k})$$

which is clearly parallel to $\mathbf{n}$. Also

$$\mathbf{m} = \frac{1}{2}(\mathbf{x}' + \mathbf{x}) = \frac{10}{9}\mathbf{i} + \frac{17}{9}\mathbf{j} + \frac{23}{9}\mathbf{k}.$$

These coordinates satisfy the equation for the plane, and so we have a double check that our image point is correctly found.

EXERCISE I

Find the image of the point $(2, 0, -1)$ under reflection in the plane $x + y + z = 3$. Check your result by the methods shown in the previous example.

The theory works equally well for $\mathbb{R}^2$. Although we generally consider reflection in a *line l* in $\mathbb{R}^2$, we could think of $\mathbb{R}^2$ as the plane $z = 0$ in $\mathbb{R}^3$, and reflection in a plane parallel to **k** and containing *l*. Then equation (7.1.1) still applies.

Find the image of the point (2,3) of $\mathbb{R}^2$ after reflection in the line $y = x - 3$. Draw a diagram to check your answer.

7.2 Rotations

Again we shall begin by considering rotations in $\mathbb{R}^3$. Suppose X' is the image of X under rotation about the line *l* which passes through the point A and is parallel to the unit vector **u**. Suppose that the angle of rotation is θ in a right hand screw sense about the direction of **u** as shown in Fig 7.2. Let P be the point on *l* for which the plane PXX' is perpendicular to *l*, and let A be a given point on *l*. We shall use the convention that **a** is the position vector of A etc.

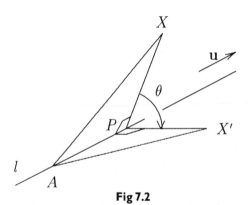

Fig 7.2

We now get considerable practice in using scalar and vector products.

$$\mathbf{AP} = ((\mathbf{x} - \mathbf{a}).\mathbf{u})\mathbf{u}$$

since **u** is a unit vector. (See Exercise 3 below.) Hence

$$\mathbf{p} = \mathbf{a} + ((\mathbf{x} - \mathbf{a}).\mathbf{u})\mathbf{u}. \tag{7.2.1}$$

For convenience, we shall set

$$\mathbf{v} = \mathbf{x} - \mathbf{p} \quad \text{and} \quad \mathbf{v}' = \mathbf{x}' - \mathbf{p}. \tag{7.2.2}$$

Then

$$\mathbf{v}' = \cos\theta\,\mathbf{v} + \sin\theta\,(\mathbf{u} \times \mathbf{v}). \tag{7.2.3}$$

(See Exercise 4 below.) Now substituting from equation (7.2.2) into (7.2.3) we get

$$\mathbf{x}' = \mathbf{p} + \cos\theta(\mathbf{x} - \mathbf{p}) + \sin\theta\{\mathbf{u} \times (\mathbf{x} - \mathbf{p})\} \tag{7.2.4}$$

where **p** is the vector defined in equation (7.2.1). This turns out to be

$$\mathbf{x}' = \mathbf{a} + ((\mathbf{x} - \mathbf{a}).\mathbf{u})\mathbf{u} + \cos\theta\{\mathbf{u} \times ((\mathbf{x} - \mathbf{a}) \times \mathbf{u})\} + \sin\theta(\mathbf{u} \times (\mathbf{x} - \mathbf{a}))$$

$$(7.2.5)$$

It is left to the reader to verify this. (See Exercise 9 in the end-of-chapter exercises.)

EXERCISE 3

With the notation above, since P lies on l, then for some $\lambda \in \mathbb{R}$,

$$\mathbf{p} = \mathbf{a} + \lambda\mathbf{u}.$$

By considering the fact that, since XP is perpendicular to l,

$$(\mathbf{x} - \mathbf{p}).\mathbf{u} = 0$$

find the value of λ in terms of $\mathbf{x}$, $\mathbf{a}$ and $\mathbf{u}$.

EXERCISE 4

With the notation above, suppose that $PX = PX' = r$:

(i) find $\mathbf{v}.\mathbf{v}'$ in terms of r and θ,
(ii) find $\mathbf{v} \times \mathbf{v}'$ in terms of r, θ and $\mathbf{u}$, and
(iii) by taking the vector product of $\mathbf{v}$ with both sides of your equation in (ii), and by using the formula for a triple vector product, show that

$$\mathbf{v}' = \cos\theta\,\mathbf{v} + \sin\theta\,(\mathbf{u} \times \mathbf{v}).$$

Example 2

Find the image of the point (1,2,3) under rotation through $\pi/3$ about the line l passing through the point $(1, 1, -1)$ and parallel to the position vector of $(1, 0, -1)$. (The angle is taken in a right hand screw sense about this vector.)

SOLUTION

$$\mathbf{a} = \mathbf{i} + \mathbf{j} - \mathbf{k}, \quad \mathbf{x} = \mathbf{i} + 2\mathbf{j} + 3\mathbf{k}, \quad \mathbf{u} = \frac{1}{\sqrt{2}}\mathbf{i} - \frac{1}{\sqrt{2}}\mathbf{k}.$$

Now

$$\mathbf{x} - \mathbf{a} = \mathbf{j} + 4\mathbf{k} \quad \text{and} \quad (\mathbf{x} - \mathbf{a}).\mathbf{u} = -2\sqrt{2}.$$

Hence

$$((\mathbf{x} - \mathbf{a}).\mathbf{u})\mathbf{u} = -2\mathbf{i} + 2\mathbf{k} \quad \text{and} \quad \mathbf{p} = -\mathbf{i} + \mathbf{j} + \mathbf{k}$$

so that

$$\mathbf{x} - \mathbf{p} = 2\mathbf{i} + \mathbf{j} + 2\mathbf{k} \quad \text{and} \quad \mathbf{u} \times (\mathbf{x} - \mathbf{p}) = \frac{1}{\sqrt{2}}\mathbf{i} - 2\sqrt{2}\mathbf{j} + \frac{1}{\sqrt{2}}\mathbf{k}$$

so substituting in (7.2.4), remembering that $\cos \pi/3 = 1/2$ and $\sin \pi/3 = \sqrt{3}/2$, we get

$$\mathbf{x}' = -\mathbf{i} + \mathbf{j} + \mathbf{k} + \frac{1}{2}(2\mathbf{i} + \mathbf{j} + 2\mathbf{k}) + \frac{\sqrt{3}}{2}\left(\frac{1}{\sqrt{2}}\mathbf{i} - 2\sqrt{2}\mathbf{j} + \frac{1}{\sqrt{2}}\mathbf{k}\right)$$

$$= \frac{\sqrt{6}}{4}\mathbf{i} + \left(\frac{3}{2} - \sqrt{6}\right)\mathbf{j} + \left(2 + \frac{\sqrt{6}}{4}\right)\mathbf{k}.$$

In $\mathbb{R}^2$ we think of rotation as rotation about a point P, but again considering $\mathbb{R}^2$ as the plane $z = 0$ in $\mathbb{R}^3$, this would be equivalent to rotation about a line through the point P parallel to $\mathbf{k}$ (that is, perpendicular to the plane). Instead of $\mathbf{u} \times \mathbf{v}$, we now think of a vector $\mathbf{v}^{\perp}$, where

$$\mathbf{v}^{\perp} = -b\mathbf{i} + a\mathbf{j} \quad \text{whenever} \quad \mathbf{v} = a\mathbf{i} + b\mathbf{j}.$$

(See Exercise 7 in the end-of-chapter exercises.) The points A and P in the above working now coincide as the centre of rotation and equation (7.2.4) is the most useful version of the required formula, with $(\mathbf{x} - \mathbf{p})^{\perp}$ replacing $\mathbf{u} \times (\mathbf{x} - \mathbf{p})$.

Example 3

Find the image of the point $(2,3)$ in the plane $\mathbb{R}^2$ under a rotation through $45°$ about the point $(1,1)$.

SOLUTION

$$\mathbf{x} = 2\mathbf{i} + 3\mathbf{j} \quad \text{and} \quad \mathbf{p} = \mathbf{i} + \mathbf{j}$$

so that

$$\mathbf{x} - \mathbf{p} = \mathbf{i} + 2\mathbf{j} \quad \text{and} \quad (\mathbf{x} - \mathbf{p})^{\perp} = -2\mathbf{i} + \mathbf{j}.$$

Therefore

$$\mathbf{x}' = \mathbf{i} + \mathbf{j} + \frac{1}{\sqrt{2}}(\mathbf{i} + 2\mathbf{j}) + \frac{1}{\sqrt{2}}(-2\mathbf{i} + \mathbf{j})$$

$$= \left(1 - \frac{1}{\sqrt{2}}\right)\mathbf{i} + \left(1 + \frac{3}{\sqrt{2}}\right)\mathbf{j}.$$

Remembering that $1/\sqrt{2}$ is approximately $2/3$ draw a diagram of X, P and X' to check that this answer is reasonable.

7.3 Translations

Translations are very easy to deal with compared with the complications of reflections and rotations. We simply add a vector. If the whole plane is translated then each point is moved by the same amount in the same direction, and so the same vector is added to the position vector of each point. Thus if X' is the image of X under translation taking the origin to the point A, then

$$\mathbf{x}' = \mathbf{x} + \mathbf{a}.$$

7.4 Isometries

Why have we considered these three types of transformation of $\mathbb{R}^3$? It is because under such transformations that distances and angles are invariant. A reflection or a rotation or a translation does not change the *size* or *shape* of a body, and the invariance of size and shape depends upon the invariance of distances and angles.

● *Definition I*

An *isometry* of $\mathbb{R}^n$ is a function $t : \mathbb{R}^n \to \mathbb{R}^n$ for which for all $\mathbf{x}, \mathbf{y} \in \mathbb{R}^n$,

$$|t(\mathbf{x}) - t(\mathbf{y})| = |\mathbf{x} - \mathbf{y}|.$$

● *Theorem I*

If $t : \mathbb{R}^n \to \mathbb{R}^n$ is an isometry, then t also preserves angles.

PROOF

Let P, Q, R be any three distinct points in $\mathbb{R}^n$ such that P', Q' and R' are their respective images under t. Then $P'Q' = PQ$, $Q'R' = QR$ and $R'P' = RP$, and by the cosine rule

$$\cos P'Q'R' = \frac{(P'Q')^2 + (R'Q')^2 - (P'R')^2}{2P'Q'.R'Q'}$$

$$= \frac{PQ^2 + RQ^2 - PR^2}{2PQ.RQ} = \cos PQR.$$

Hence angles are preserved under t.

Note that it is the *size* of the angle which is preserved. In a reflection for example, angles are reversed, but since $\cos \theta = \cos(-\theta)$, the above equation is still satisfied if angles are reversed.

● *Theorem 2*

Translations, reflections and rotations are isometries of $\mathbb{R}^3$.

PROOF
(i) Consider the translation $t : \mathbb{R}^3 \to \mathbb{R}^3$ given by

$$t(\mathbf{x}) = \mathbf{x} + \mathbf{a}.$$

Then

$$|t(\mathbf{x}) - t(\mathbf{y})| = |(\mathbf{x} + \mathbf{a}) - (\mathbf{y} + \mathbf{a})| = |\mathbf{x} - \mathbf{y}|$$

and the condition is satisfied. Hence a translation is an isometry.

(ii) Let $t : \mathbb{R}^3 \to \mathbb{R}^3$ be a reflection. Then

$$|t(\mathbf{x}) - t(\mathbf{y})| = |(\mathbf{x} - \mathbf{y}) - 2((\mathbf{x} - \mathbf{y}).\mathbf{n})\mathbf{n}|$$

Now

$$|t(\mathbf{x}) - t(\mathbf{y})|^2 = \{t(\mathbf{x}) - t(\mathbf{y})\} \cdot \{t(\mathbf{x}) - t(\mathbf{y})\}$$

$$= (\mathbf{x} - \mathbf{y}) \cdot (\mathbf{x} - \mathbf{y}) - 4(\mathbf{x} - \mathbf{y}) \cdot \{((\mathbf{x} - \mathbf{y}) \cdot \mathbf{n})\mathbf{n}\} + 4((\mathbf{x} - \mathbf{y}) \cdot \mathbf{n})^2 \mathbf{n} \cdot \mathbf{n}$$

$$= |\mathbf{x} - \mathbf{y}|^2 - 4|(\mathbf{x} - \mathbf{y}) \cdot \mathbf{n}|^2 + 4|(\mathbf{x} - \mathbf{y}) \cdot \mathbf{n}|^2$$

$$= |\mathbf{x} - \mathbf{y}|^2.$$

Hence

$$|t(\mathbf{x}) - t(\mathbf{y})| = |\mathbf{x} - \mathbf{y}|$$

and a reflection is an isometry.

(iii) See Exercise 5 below.

EXERCISE 5

(i) Show that with the notation of this section

$$\mathbf{x}' - \mathbf{y}' = \Big((\mathbf{x} - \mathbf{y}) \cdot \mathbf{u}\Big)\mathbf{u} + \cos\theta \Big\{\mathbf{u} \times \big((\mathbf{x} - \mathbf{y}) \times \mathbf{u}\big)\Big\} + \sin\theta\Big(\mathbf{u} \times (\mathbf{x} - \mathbf{y})\Big).$$

(ii) Explain why $\mathbf{u}$, $\mathbf{u} \times (\mathbf{x} - \mathbf{y})$, and $\mathbf{u} \times ((\mathbf{x} - \mathbf{y}) \times \mathbf{u})$ are mutually orthogonal.

(iii) If $\mathbf{a}$, $\mathbf{b}$ and $\mathbf{c}$ are mutually orthogonal, and if

$$\mathbf{v} = \alpha\mathbf{a} + \beta\mathbf{b} + \gamma\mathbf{c},$$

show that

$$\mathbf{v} \cdot \mathbf{v} = \alpha^2(\mathbf{a} \cdot \mathbf{a}) + \beta^2(\mathbf{b} \cdot \mathbf{b}) + \gamma^2(\mathbf{c} \cdot \mathbf{c}).$$

(iv) By putting $\mathbf{a} = ((\mathbf{x} - \mathbf{y}) \cdot \mathbf{u})\mathbf{u}$, $\mathbf{b} = \mathbf{u} \times ((\mathbf{x} - \mathbf{y}) \times \mathbf{u})$, and $\mathbf{c} = \mathbf{u} \times (\mathbf{x} - \mathbf{y})$, show that $\mathbf{a} \cdot \mathbf{a} = ((\mathbf{x} - \mathbf{y}) \cdot \mathbf{u})^2$, and $\mathbf{b} \cdot \mathbf{b} = \mathbf{c} \cdot \mathbf{c} = (\mathbf{x} - \mathbf{y}) \cdot (\mathbf{x} - \mathbf{y}) - (\mathbf{u} \cdot (\mathbf{x} - \mathbf{y}))^2$.

(v) Use the above results to show that

$$(\mathbf{x}' - \mathbf{y}') \cdot (\mathbf{x}' - \mathbf{y}') = (\mathbf{x} - \mathbf{y}) \cdot (\mathbf{x} - \mathbf{y}),$$

and hence that a rotation of $\mathbb{R}^3$ is an isometry.

7.5 Combinations of reflections, rotations and translations

It is not too difficult to see that if we have a translation through $\mathbf{a}$ followed by a translation through $\mathbf{b}$, the resultant is a translation through $\mathbf{a} + \mathbf{b}$, since if

$$\mathbf{x}' = \mathbf{x} + \mathbf{a} \quad \text{and} \quad \mathbf{x}'' = \mathbf{x}' + \mathbf{b}$$

then

$$\mathbf{x}'' = (\mathbf{x} + \mathbf{a}) + \mathbf{b} = \mathbf{x} + (\mathbf{a} + \mathbf{b}).$$

What happens when a reflection is followed by a reflection? This depends upon the relationship between the two planes.

Case (i) Successive Reflection in Two Parallel Planes

Consider the two planes π and π' whose equations are

$$(\mathbf{r} - \mathbf{a}).\mathbf{n} = 0 \quad\text{and}\quad (\mathbf{r} - \mathbf{b}).\mathbf{n} = 0.$$

Suppose reflection in π takes X to X', and reflection in π' takes X' to X''. Then

$$\mathbf{x}' = \mathbf{x} - 2((\mathbf{x} - \mathbf{a}).\mathbf{n})\mathbf{n} \quad\text{and}\quad \mathbf{x}'' = \mathbf{x}' - 2((\mathbf{x}' - \mathbf{b}).\mathbf{n})\mathbf{n}.$$

Thus

$$\begin{aligned}
\mathbf{x}'' &= \mathbf{x} - 2((\mathbf{x} - \mathbf{a}).\mathbf{n})\mathbf{n} - 2[\{\mathbf{x} - 2((\mathbf{x} - \mathbf{a}).\mathbf{n})\mathbf{n} - \mathbf{b}\}.\mathbf{n}]\mathbf{n} \\
&= \mathbf{x} - 2((\mathbf{a} - \mathbf{b}).\mathbf{n})\mathbf{n}.
\end{aligned}$$

This is a translation by $2d\mathbf{n}$, where d is the distance between the parallel planes.

So, a reflection in one plane followed by reflection in a parallel plane is a translation through a distance which is twice the distance between the two planes, and in a direction perpendicular to both planes.

Case (ii) Successive Reflection in Two Non-parallel Planes

It is also possible to show that a reflection in a plane π followed by a reflection in a plane π', where π and π' are not parallel, is a rotation about the line of intersection of the two planes, through an angle which is twice the angle between the two planes. This is quite complicated to work out, but for those interested, it is given as Exercise 13 in the end-of-chapter exercises, with hints to guide you through the working.

So, we have found that both translations and rotations can be thought of as combinations of reflections, and it is true to say that all isometries of $\mathbb{R}^3$ are generated by the reflections in $\mathbb{R}^3$; that is to say, that every isometry of $\mathbb{R}^3$ can be effected by a combination of reflections.

All of these isometries are much easier to deal with if the planes of reflection or the axes of rotation contain axes of coordinates, and this is where the orthonormal bases from Chapter 5 are useful. For example, if we want the reflection of the vector $a\mathbf{i} + b\mathbf{j} + c\mathbf{k}$ in the plane $z = 0$ (that is, the plane containing the x- and y-axes), it is simply the vector $a\mathbf{i} + b\mathbf{j} - c\mathbf{k}$. Likewise, if $\{\mathbf{u}, \mathbf{v}, \mathbf{w}\}$ is an orthonormal basis of $\mathbb{R}^3$, the reflection of a point whose position vector is $\alpha\mathbf{u} + \beta\mathbf{v} + \gamma\mathbf{w}$ in the plane $\mathbf{r}.\mathbf{w} = 0$ is the point whose position vector is $\alpha\mathbf{u} + \beta\mathbf{v} - \gamma\mathbf{w}$. Similar results are true for rotations. A method of changing bases, and in particular changing from one orthonormal basis to another, will be found in the linear algebra book in this series.

Summary

1. For a reflection, if $\mathbf{a}$ is the position vector of a point on the plane of reflection,

$$\mathbf{x}' = \mathbf{x} - 2((\mathbf{x} - \mathbf{a}).\mathbf{n})\mathbf{n}. \tag{7.1.1}$$

2. For a rotation, if **a** is the position vector of a point on the axis of rotation, and **u** is a unit vector parallel to this axis,

$$\mathbf{x}' = \mathbf{a} + ((\mathbf{x} - \mathbf{a}).\mathbf{u})\mathbf{u} + \cos\theta\{\mathbf{u} \times ((\mathbf{x} - \mathbf{a}) \times \mathbf{u})\} + \sin\theta(\mathbf{u} \times (\mathbf{x} - \mathbf{a})).$$

$$(7.2.5)$$

3. A translation is given by $\mathbf{x}' = \mathbf{x} + \mathbf{a}$.
4. An isometry of $\mathbb{R}^n$ is a function $t : \mathbb{R}^n \to \mathbb{R}^n$ such that

$$\forall \mathbf{x}, \mathbf{y} \in \mathbb{R}^n, \quad |t(\mathbf{x}) - t(\mathbf{y})| = |\mathbf{x} - \mathbf{y}|.$$

5. Reflections, rotations and translations or any combination of these are isometries.
6. Any isometry of $\mathbb{R}^3$ can be regarded as a combination of reflections.

FURTHER EXERCISES

6. Suppose $\mathbf{a} = \mathbf{i} + 2\mathbf{j} + 3\mathbf{k}$ and $\mathbf{u} = 2\mathbf{i} - \mathbf{j} + 2\mathbf{k}$, and let $\mathbf{b} = \mathbf{a} \times \mathbf{u}$ and $\mathbf{c} = \mathbf{u} \times (\mathbf{a} \times \mathbf{u})$.
   (i) Find **b** and **c** in terms of **i**, **j** and **k**.
   (ii) Find $\mathbf{u}.\mathbf{b}$, $\mathbf{u}.\mathbf{c}$ and $\mathbf{b}.\mathbf{c}$.
   (iii) What do the results of (ii) tell us about **u**, **b** and **c**? Could this have been deduced from the definition of the vector product?
7. (i) If, in $\mathbb{R}^3$, $\mathbf{v} = a\mathbf{i} + b\mathbf{j}$, find $\mathbf{k} \times \mathbf{v}$.
   (ii) If, in $\mathbb{R}^2$, $\mathbf{v} = a\mathbf{i} + b\mathbf{j}$, find $\mathbf{v}^\perp$, and compare your answer with that of (i).
8. Find the image of the point $(2,0,0)$ under rotation through π about the line though the point $(1,1,1)$ parallel to the x-axis.
9. (Note: This is referred to at the end of Section 7.2.) Given that $\mathbf{p} = \mathbf{a} + ((\mathbf{x} - \mathbf{a}).\mathbf{u})\mathbf{u}$, where **u** is a unit vector,
   (i) show that $\mathbf{x} - \mathbf{p} = \mathbf{u} \times ((\mathbf{x} - \mathbf{a}) \times \mathbf{u})$,
   (ii) show that $\mathbf{u} \times (\mathbf{x} - \mathbf{p}) = \mathbf{u} \times (\mathbf{x} - \mathbf{a})$, and
   (iii) by using the results of (i) and (ii) above in equation (7.2.4) derive equation (7.2.5).
10. (i) Find the image of the point (p, q, r) under
       (a) reflection in the plane $x = 0$,
       (b) reflection in the plane $y = 0$,
       (c) reflection in the plane $x = 0$ followed by reflection in the plane $y = 0$.
    (ii) Find the image of the point (p, q, r) under a rotation of $180°$ about the z-axis.
    (iii) Compare your answers with (i) and (ii). Explain this result.
11. Show that the linear transformation of $\mathbb{R}^3$ whose matrix is

$$\begin{pmatrix} 1/3 & 2/3 & 2/3 \\ 2/3 & 1/3 & -2/3 \\ 2/3 & -2/3 & 1/3 \end{pmatrix}$$

is an isometry. (*Hint:* For any vectors $\mathbf{x}, \mathbf{y} \in \mathbb{R}^3$, let $\mathbf{x} - \mathbf{y} = \mathbf{z} = \begin{pmatrix} p \\ q \\ r \end{pmatrix}$, and show that $A\mathbf{z}.A\mathbf{z} = \mathbf{z}.\mathbf{z}$.)

12. (i) If, in equation (7.1.1), $\mathbf{a} = \mathbf{0}$, the reflection is a linear transformation, and equation (7.1.1) can therefore be written in the form

$$\mathbf{x}' = A\mathbf{x}.$$

What is the matrix A in this case, given that the unit vector $\mathbf{n} = \begin{pmatrix} n_1 \\ n_2 \\ n_3 \end{pmatrix}$?

(ii) If, in equation (7.2.5), $\mathbf{a} = \mathbf{0}$, the reflection is a linear transformation, and equation (7.2.5) can therefore be written in the form

$$\mathbf{x}' = A\mathbf{x}.$$

What is the matrix A in this case, given that the unit vector $\mathbf{u} = \begin{pmatrix} u_1 \\ u_2 \\ u_3 \end{pmatrix}$?

13. CHALLENGE QUESTION   In this question we show that successive reflection in two non-parallel planes π and π' is equivalent to a rotation about the line of intersection of π and π' through an angle 2θ, where θ is the angle between the two planes. We simplify matters by choosing our origin to lie on the line l of intersection, and therefore the equations of π and π' are respectively

$$\mathbf{r}.\mathbf{m} = 0 \quad \text{and} \quad \mathbf{r}.\mathbf{n} = 0,$$

where $\mathbf{m}$ and $\mathbf{n}$ are non-parallel unit vectors.

(i) If $\mathbf{u}$ is a unit vector in the direction of l explain why
   (a) $\mathbf{m} \times \mathbf{n} = (\sin\theta)\mathbf{u}$, and
   (b) $\mathbf{m}.\mathbf{n} = \cos\theta$.

(ii) Show that if $\mathbf{x}'$ is the image of $\mathbf{x}$ after reflection in π', and $\mathbf{x}''$ is the image of $\mathbf{x}'$ after reflection in π, then

$$\mathbf{x}'' = \mathbf{x} - 2(\mathbf{x}.\mathbf{n})\mathbf{n} - 2(\mathbf{x}.\mathbf{m})\mathbf{m} + 4(\mathbf{x}.\mathbf{n})(\mathbf{n}.\mathbf{m})\mathbf{m}.$$

(iii) Starting from equation (7.2.5), by putting $\mathbf{a} = \mathbf{0}$, by replacing θ by 2θ, and then by substituting from (i) derive the RHS of the equation given in (ii). (*Hint*: You will need to use the expansion of a triple vector product, sometimes more than once!)

8 • Vector-valued Functions of a Single Variable

8.1 Parameters

Graphs in the plane can be regarded as vector-valued functions, even when they are given in the form $y = f(x)$, for both x and y are functions of x. Thus any point on the curve whose equation is $y = f(x)$ has position vector $x\mathbf{i} + f(x)\mathbf{j}$. For example, the position vector of any point on the parabola whose equation is $y = x^2$ is of the form $x\mathbf{i} + x^2\mathbf{j}$. It is more practical to use the vectors $\mathbf{i}$ and $\mathbf{j}$ here, since we may have quite complicated functions involved, and writing these as column vectors might be somewhat messy. Alternatively a curve in $\mathbb{R}^2$ can be given in terms of a parameter different from x. Most readers will be familiar with the parametric definition of the parabola $y^2 = 4ax$ as $x = at^2$, $y = 2at$. Here a point on the curve has position vector $at^2\mathbf{i} + 2at\mathbf{j}$.

It is sometimes convenient to think of the parameter as indicating time, so as a stopwatch ticks from its zero position in units of time (seconds, minutes, hours, …) so we can think of a *blob* travelling along one half of the parabola, and, as the watch ticks on and on, so the blob will move further and further along the curve away from the vertex. If we allow that the watch was running even before we started recording, then the blob would have been travelling along the parabola in the very same way through negative values of t until the time $t = 0$ had occurred and the vertex had been arrived at. In this way, as $t \to +\infty$ the corresponding point travels off to infinity along the upper half of the curve, and as $t \to -\infty$ the point travels off to infinity along the lower half of the curve. In this example for each point on the curve there is exactly one value of t which is its parameter. Points on the curve given by the values $t = 0$, $t = \pm1$, $t = \pm2$ are shown in Fig 8.1.

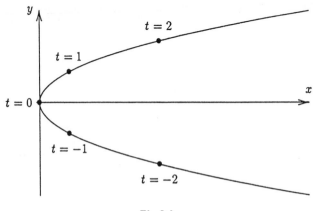

Fig 8.1

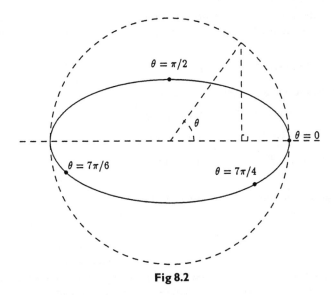

Fig 8.2

Of course, time is not the only parameter we could consider. If we think of the point given by the parametric equations

$$x = a\cos\theta \qquad y = b\sin\theta$$

for an ellipse, then we think of θ as being the angle shown in Fig 8.2. In this case, although each value of θ corresponds to only one point on the curve, each point on the curve corresponds to an infinite number of values of θ, as the parameter $\theta + 2n\pi$ for any $n \in \mathbf{Z}$ will give the same point on the curve as the parameter θ.

Once again, particular values of θ are shown on the diagram.

EXERCISE I

Draw the curve in the plane whose general point has position vector

$$\mathbf{r} = (2\sin t)\mathbf{i} + (2\cos t)\mathbf{j},$$

marking the points where $t = 0$, $t = \pi/2$, $t = -\pi/3$, $t = \pi$.

One of the most useful parameters is the arc length measured from a fixed point on the given curve, and we shall make use of this parameter later in the chapter.

8.2 Differentiation of vectors and derived vectors in $\mathbb{R}^2$

In this chapter we shall assume, unless otherwise stated, that our functions f, g and h are continuous and continuously differentiable at least as far as the second derivative.

Consider a curve C given in terms of a parameter t by the vector equation

$$\mathbf{r} = f(t)\mathbf{i} + g(t)\mathbf{j} \tag{8.2.1}$$

and consider the point P on C whose parameter is t. Suppose we increase t by a small amount δt; then the point Q with this parameter will have position vector $\mathbf{r} + \delta\mathbf{r}$, where

$$\mathbf{r} + \delta\mathbf{r} = f(t + \delta t)\mathbf{i} + g(t + \delta t)\mathbf{j}. \tag{8.2.2}$$

Subtracting equation (8.2.1) from (8.2.2) we get

$$\delta\mathbf{r} = \{\,f(t + \delta t) - f(t)\}\mathbf{i} + \{g(t + \delta t) - g(t)\}\mathbf{j}.$$

Dividing both sides of this equation by δt we get

$$\frac{\delta\mathbf{r}}{\delta t} = \frac{f(t + \delta t) - f(t)}{\delta t}\,\mathbf{i} + \frac{g(t + \delta t) - g(t)}{\delta t}\,\mathbf{j},$$

and as we allow δt to tend to zero, the right hand side tends to

$$\frac{df}{dt}\mathbf{i} + \frac{dg}{dt}\mathbf{j}.$$

But we see from Fig 8.3 that, as we allow δt to tend to zero, the vector $\delta\mathbf{r}$ approaches the direction of the tangent and $\delta\mathbf{r}/\delta t$ will approach $d\mathbf{r}/dt$, the *rate of change of position vector*, which is, in essence, a *velocity vector*.

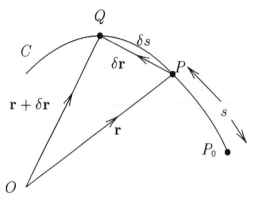

Fig 8.3

● *Definition I*

If a position vector is given in terms of a parameter α, so that

$$\mathbf{r} = f(\alpha)\mathbf{i} + g(\alpha)\mathbf{j},$$

then the vector $d\mathbf{r}/d\alpha$ given by

$$\frac{d\mathbf{r}}{d\alpha} = \frac{df}{d\alpha}\mathbf{i} + \frac{dg}{d\alpha}\mathbf{j}$$

is called the *derived vector* or *tangent vector* of $\mathbf{r}$ with respect to α.

There are two parameters for which the derivatives have standard abbreviations. The first is the case where the parameter t represents time, and we write

$$\dot{f} = \frac{df}{dt} \quad \text{and} \quad \dot{g} = \frac{dg}{dt},$$

and the second case is differentiation with respect to x, where we write

$$f'(x) = \frac{df}{dx} \quad \text{and} \quad g'(x) = \frac{dg}{dx}.$$

We used the term *tangent vector* above. As in ordinary calculus, derivatives help us to find tangents. Suppose we consider our parabola again, and let us take a particular value of a, say $a = 1/2$. Then a general point on this parabola has position vector

$$\mathbf{r} = \frac{1}{2}t^2\mathbf{i} + t\mathbf{j}.$$

Now if we differentiate this vector with respect to t component-wise we get the derived vector

$$\dot{\mathbf{r}} = t\mathbf{i} + \mathbf{j},$$

and if we draw the direction of the above vector at the point whose parameter is t on the diagram in Fig 8.1, we see that the derived vector is parallel to the tangent to the curve at that point. By considering the particular values of t we picked out before, we confirm this tangency property by drawing the derived vectors at these points. The magnitudes (given by $\sqrt{1 + t^2}$) of the derived vectors are written near the corresponding points on the curve in Fig 8.4.

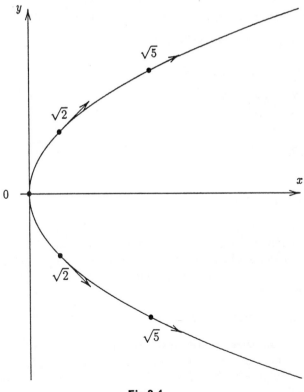

Fig 8.4

If the position vector **r** of the point P is given in terms of a time parameter t, then $\dot{\mathbf{r}}$ gives the velocity vector at this particular time, and being a vector this means in both magnitude and direction. So, in the case above, we see that as t increases from $-\infty$ to $+\infty$ the particle describing the curve continuously slows down until it reaches the $t = 0$ point, and then speeds up again ever increasingly as $t \to \infty$.

If we differentiate again, then we see that in this case

$$\ddot{\mathbf{r}} = \mathbf{i} + 0\mathbf{j} = \mathbf{i}$$

and the acceleration is seen to be constant with no component on the y-direction, and 1 unit in the positive x-direction.

EXERCISE 2

Draw a copy of Fig 8.2, calculate the tangent vector $d\mathbf{r}/d\theta$ at a general point, and, using the values of θ given in Exercise 8.1, draw in the directions of the tangent vectors at the points concerned.

8.3 Curves in three dimensions

So far in this chapter we have dealt only with curves in two dimensions, but the reasoning at the beginning of the last section can be extended to curves in $\mathbb{R}^3$, and the ideas of tangency, velocity and acceleration still apply. In three dimensions a curve can be defined parametrically by a single variable as follows:

$$\mathbf{r} = f(\alpha)\mathbf{i} + g(\alpha)\mathbf{j} + h(\alpha)\mathbf{k}$$

and the derived vector will be

$$\frac{d\mathbf{r}}{d\alpha} = \frac{df}{d\alpha}\mathbf{i} + \frac{dg}{d\alpha}\mathbf{j} + \frac{dh}{d\alpha}\mathbf{k}.$$

Example 1

Consider the curve given by

$$\mathbf{r} = a\cos t\,\mathbf{i} + a\sin t\,\mathbf{j} + bt\mathbf{k}$$

where a and b are constants.

This is the equation of a helix. If $b = 0$ the curve is simply a circle in the plane $z = 0$, and as t increases from $-\infty$ to ∞ so the point with parameter t continues to describe the circle infinitely many times. However, if $b > 0$ as t increases so does the height and since

$$\dot{\mathbf{r}} = -a\sin t\,\mathbf{i} + a\cos t\,\mathbf{j} + b\mathbf{k}$$

the point with parameter t moves upwards at constant speed on a helix described on a vertical circular cylinder of radius a as shown in Fig 8.5. If $b < 0$ then we should still obtain a helix, but it would be the reflection in the plane $z = 0$ of the helix in Fig 8.5.

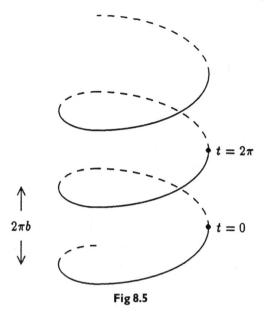

Fig 8.5

Given that

$$\mathbf{r} = t \cos t\, \mathbf{i} + t \sin t\, \mathbf{j} + t\mathbf{k}$$

find $\dot{\mathbf{r}}$ and $\ddot{\mathbf{r}}$. Describe the curve this represents, if possible with a diagram.

8.4 Rules for differentiating vectors

In differentiation of scalar functions we have the addition rule, the product rule, the quotient rule, and so on. With vectors we have an addition rule, and since there are three sorts of product involving vectors, we need a rule for each of these, but they follow the same basic principle. The most important rules for differentiating vector-valued functions $\mathbf{u}$ and $\mathbf{v}$ are given below:

Rule 1: If $\mathbf{c}$ is a constant vector, $\dfrac{d\mathbf{c}}{dt} = \mathbf{0}$.

Rule 2: $\dfrac{d}{dt}(\mathbf{u} + \mathbf{v}) = \dfrac{d\mathbf{u}}{dt} + \dfrac{d\mathbf{v}}{dt}$.

Rule 3: If ϕ is a scalar function, $\dfrac{d}{dt}(\phi\mathbf{u}) = \dfrac{d\phi}{dt}\mathbf{u} + \phi\dfrac{d\mathbf{u}}{dt}$.

Rule 4: $\dfrac{d}{dt}(\mathbf{u}.\mathbf{v}) = \mathbf{u}.\dfrac{d\mathbf{v}}{dt} + \dfrac{d\mathbf{u}}{dt}.\mathbf{v}$.

Rule 5: $\dfrac{d}{dt}(\mathbf{u} \times \mathbf{v}) = \mathbf{u} \times \dfrac{d\mathbf{v}}{dt} + \dfrac{d\mathbf{u}}{dt} \times \mathbf{v}$.

● *Example 2*

We shall show that Rule 4 above holds. Suppose

$$\mathbf{u} = f_1(t)\mathbf{i} + g_1(t)\mathbf{j} + h_1(t)\mathbf{k} \quad \text{and} \quad \mathbf{v} = f_2(t)\mathbf{i} + g_2(t)\mathbf{j} + h_2(t)\mathbf{k}.$$

Then

$$\frac{d\mathbf{u}}{dt} = \frac{df_1}{dt}\mathbf{i} + \frac{dg_1}{dt}\mathbf{j} + \frac{dh_1}{dt}\mathbf{k}$$

$$\frac{d\mathbf{v}}{dt} = \frac{df_2}{dt}\mathbf{i} + \frac{dg_2}{dt}\mathbf{j} + \frac{dh_2}{dt}\mathbf{k}$$

so that

$$\mathbf{u} \cdot \frac{d\mathbf{v}}{dt} + \frac{d\mathbf{u}}{dt} \cdot \mathbf{v} = f_1\frac{df_2}{dt} + g_1\frac{dg_2}{dt} + h_1\frac{dh_2}{dt} + \frac{df_1}{dt}f_2 + \frac{dg_1}{dt}g_2 + \frac{dh_1}{dt}h_2$$

$$= f_1\frac{df_2}{dt} + \frac{df_1}{dt}f_2 + g_1\frac{dg_2}{dt} + \frac{dg_1}{dt}g_2 + h_1\frac{dh_2}{dt} + \frac{dh_1}{dt}h_2$$

$$= \frac{d}{dt}(f_1f_2 + g_1g_2 + h_1h_2)$$

$$= \frac{d}{dt}(\mathbf{u}.\mathbf{v}).$$

EXERCISE 4

Check that the other rules above are satisfied by differentiating the components of the vectors on the LHS of each equation, and by comparing with the components on the RHS.

EXERCISE 5

If the position vector of an object is given in terms of time t by

$$\mathbf{r} = t\mathbf{i} + t^2\mathbf{j} + t^3\mathbf{k}$$

find the components of its velocity and acceleration in the direction of the vector $\mathbf{i} + \mathbf{j} - \mathbf{k}$.

8.5 The Serret–Frenet equations for a curve in $\mathbb{R}^3$

As mentioned at the end of Section 8.1 one of the useful parameters for a point P on a curve is the distance s of the point P from a fixed point P_0 *measured along the curve*. If we look back to Fig 8.3 we can see that not only does $\delta\mathbf{r}$ approach the tangent in direction as $\delta t \to 0$, but also the length $|\delta\mathbf{r}|$ approaches δs, the length of arc from P to Q. This means that in the limiting case $d\mathbf{r}/ds$ is a *unit* vector. Suppose we call this unit vector $\mathbf{t}$; then $\mathbf{t}$ is the *unit tangent vector* to the curve at P.

Since **t** is a unit vector $\mathbf{t}.\mathbf{t} = 1$, which is a constant, and hence by Rules 1 and 4,

$$0 = \frac{d}{ds}(\mathbf{t}.\mathbf{t}) = \mathbf{t}.\frac{d\mathbf{t}}{ds} + \frac{d\mathbf{t}}{ds}.\mathbf{t} = 2\mathbf{t}.\frac{d\mathbf{t}}{ds}.$$

From this equation we see that

$$\mathbf{t}.\frac{d\mathbf{t}}{ds} = 0 \tag{8.5.1}$$

and, provided $d\mathbf{t}/ds \neq \mathbf{0}$, this means that $d\mathbf{t}/ds$ is perpendicular to **t**. (We shall consider later the case where $d\mathbf{t}/ds$ is zero.)

Let **n** denote the unit vector in the direction of $d\mathbf{t}/ds$, so that

$$\frac{d\mathbf{t}}{ds} = \kappa\mathbf{n} \tag{8.5.2}$$

for some scalar function κ of s. κ is called the *curvature* of C at the point P with parameter s, and **n** is called the *unit principal normal* to the curve. In fact, if γ is a circle which touches C at P, which has the same curvature as C at P, and which lies in the plane through P parallel to both **t** and **n**, then the radius of γ will be $\rho = 1/\kappa$, and we call ρ the *radius of curvature* of C at P. Now **t** and **n** are orthogonal unit vectors, and so their vector product is a unit vector. We call this vector

$$\mathbf{b} = \mathbf{t} \times \mathbf{n}$$

the unit *binormal* of C at P.

Because **t**, **n** and **b** are all unit vectors which are mutually orthogonal, by the arguments leading to equation (8.5.1) we have

$$\mathbf{t}.\frac{d\mathbf{t}}{ds} = \mathbf{n}.\frac{d\mathbf{n}}{ds} = \mathbf{b}.\frac{d\mathbf{b}}{ds} = 0, \tag{8.5.3}$$

and, because of mutual orthogonality,

$$\mathbf{t}.\mathbf{n} = \mathbf{n}.\mathbf{b} = \mathbf{b}.\mathbf{t} = 0. \tag{8.5.4}$$

Now, using Rule 5,

$$\frac{d\mathbf{b}}{ds} = \mathbf{t} \times \frac{d\mathbf{n}}{ds} + \frac{d\mathbf{t}}{ds} \times \mathbf{n} = \mathbf{t} \times \frac{d\mathbf{n}}{ds}$$

since **n** is parallel to $d\mathbf{t}/ds$, and hence their vector product is zero. Taking the vector product of **n** with each side of this equation

$$\mathbf{n} \times \frac{d\mathbf{b}}{ds} = \mathbf{n} \times \left(\mathbf{t} \times \frac{d\mathbf{n}}{ds}\right) = \left(\mathbf{n}.\frac{d\mathbf{n}}{ds}\right)\mathbf{t} - (\mathbf{n}.\mathbf{t})\frac{d\mathbf{n}}{ds} = \mathbf{0}$$

by the results of equations (8.5.3) and (8.5.4).

This means that, provided $d\mathbf{b}/ds \neq \mathbf{0}$, **n** and $d\mathbf{b}/ds$ must be parallel, that is

$$\frac{d\mathbf{b}}{ds} = -\tau\mathbf{n} \tag{8.5.5}$$

for some scalar function τ of s. We call τ the *torsion* of C at P.

Since **t**, **n**, **b** are mutually orthogonal unit vectors, and since $\mathbf{b} = \mathbf{t} \times \mathbf{n}$, we can deduce that $\mathbf{n} = \mathbf{b} \times \mathbf{t}$, and by differentiating both sides of this we obtain

$$\frac{d\mathbf{n}}{ds} = \mathbf{b} \times \frac{d\mathbf{t}}{ds} + \frac{d\mathbf{b}}{ds} \times \mathbf{t} = \kappa \mathbf{b} \times \mathbf{n} - \tau \mathbf{n} \times \mathbf{t},$$

that is

$$\frac{d\mathbf{n}}{ds} = \tau \mathbf{b} - \kappa \mathbf{t}. \tag{8.5.6}$$

Collecting together equations (8.5.2), (8.5.5) and (8.5.6) we have the three equations

$$\frac{d\mathbf{t}}{ds} = \kappa \mathbf{n}$$

$$\frac{d\mathbf{n}}{ds} = \tau \mathbf{b} - \kappa \mathbf{t} \tag{$*$}$$

$$\frac{d\mathbf{b}}{ds} = -\tau \mathbf{n}$$

and these are called the *Serret–Frenet equations*.

If the curvature κ is large at a point P, then our curve is very bent near that point, and if κ is small, then the curve approaches straightness near P. If $\kappa = 0$ at a point P, then we say that the curve is *straight* at P. This does not mean that the curve is actually in a straight line close to P, although that is one possibility, but P may be at a point where the curve bends one way on one side of P and in the opposite way on the other side of P. We can compare this with a point of inflection in calculus, where the second derivative of the function is zero, since if $\kappa = 0$, from the Serret–Frenet equations,

$$0 = \frac{d\mathbf{t}}{ds} = \frac{d^2\mathbf{r}}{ds^2}.$$

If $\kappa = 0$ for the whole curve, then the curve is a straight line throughout.

What happens if $\tau = 0$? In this case **b** is stationary at the point, so there is no *twist* in the curve at that point, and if $\tau = 0$ for the whole curve, then the curve lies entirely within a plane. So the curvature κ measures how *bent* the curve is, and the torsion τ measures how *twisted* the curve is.

Going back to our example of the helix defined by

$$\mathbf{r} = (a\cos\theta)\mathbf{i} + (a\sin\theta)\mathbf{j} + b\theta\mathbf{k}, \tag{8.5.7}$$

we find that

$$\frac{d\mathbf{r}}{ds} = \frac{d\mathbf{r}}{d\theta}\frac{d\theta}{ds} = \{(-a\sin\theta)\mathbf{i} + (a\cos\theta)\mathbf{j} + b\mathbf{k}\}\frac{d\theta}{ds}.$$

Now we know that $d\mathbf{r}/ds$ is a unit vector, so that

$$\sqrt{a^2\sin^2\theta + a^2\cos^2\theta + b^2}\,\frac{d\theta}{ds} = 1$$

and hence

$$\frac{d\theta}{ds} = \frac{1}{\sqrt{a^2 + b^2}} = c, \quad \text{say.}$$

Now $\mathbf{t} = d\mathbf{r}/ds$, so that

$$\mathbf{t} = c\{(-a\sin\theta)\mathbf{i} + (a\cos\theta)\mathbf{j} + b\mathbf{k}\}$$

which means that

$$\kappa\mathbf{n} = \frac{d\mathbf{t}}{ds} = -ca\{(\cos\theta)\mathbf{i} + (\sin\theta)\mathbf{j}\}\frac{d\theta}{ds}$$

and since $\mathbf{n}$ is a unit vector, $\kappa = c^2 a$ which leads to

$$\kappa = \frac{a}{a^2 + b^2} \quad \text{and} \quad \mathbf{n} = (-\cos\theta)\mathbf{i} + (-\sin\theta)\mathbf{j}.$$

We know that $\mathbf{b} = \mathbf{t} \times \mathbf{n}$, so by differentiating this (see Exercise 6 below) we get

$$\frac{d\mathbf{b}}{ds} = \frac{b}{a^2 + b^2}\{(\cos\theta)\mathbf{i} + (\sin\theta)\mathbf{j}\} = -\frac{b}{a^2 + b^2}\mathbf{n} \qquad (8.5.8)$$

and from the third Serret–Frenet equation, this tells us that

$$\tau = \frac{b}{a^2 + b^2}.$$

Thus for the helix whose equation is given in equation (8.5.7) we have found that both the curvature and the torsion are constant at every point.

EXERCISE 6

In the above example we have found $\mathbf{t}$ and $\mathbf{n}$ in terms of $\mathbf{i}$, $\mathbf{j}$ and $\mathbf{k}$. Use these to express $\mathbf{b}$ in terms of these standard unit vectors, and then differentiate to obtain equation (8.5.8). (*Reminder*: Don't forget the $d\theta/ds$ at each stage of differentiation.)

EXERCISE 7

If $b = 0$ in the above example, what are the values of κ and τ, and why should you not be surprised by this?

Summary

1. Curves in $\mathbb{R}^3$ can be defined parametrically by a single variable as follows:

$$\mathbf{r} = f(\alpha)\mathbf{i} + g(\alpha)\mathbf{j} + h(\alpha)\mathbf{k}$$

(in $\mathbb{R}^2$ we simply have $h(\alpha) = 0$).
2. The derived vector of $\mathbf{r}$ above with respect to α will be

$$\frac{d\mathbf{r}}{d\alpha} = \frac{df}{d\alpha}\mathbf{i} + \frac{dg}{d\alpha}\mathbf{j} + \frac{dh}{d\alpha}\mathbf{k}.$$

3. Rules for differentiating sums and products follow the usual pattern for derivatives of sums and products of vector-valued functions, and these can be found in Section 8.4.
4. The Serret–Frenet equations involve the unit tangent, normal and binormal vectors to the curve, **t**, **n** and **b** respectively, and are

$$\frac{d\mathbf{t}}{ds} = \kappa\mathbf{n}$$

$$\frac{d\mathbf{n}}{ds} = \tau\mathbf{b} - \kappa\mathbf{t} \tag{$*$}$$

$$\frac{d\mathbf{b}}{ds} = -\tau\mathbf{n}$$

where s measures the arc length from a given point on the curve.
5. In 4 above, κ is called the *curvature*, τ is called the *torsion*, and $\rho = 1/\kappa$ is called the *radius of curvature* of the curve at the point concerned.

FURTHER EXERCISES

8. Draw the graph of the vector-valued function

$$\mathbf{r} = \cos^3\theta\,\mathbf{i} + \sin^3\theta\,\mathbf{j},$$

and find the tangent vector at the values

$$\theta = 0, \ \pm\pi/6, \ \pm\pi/4, \ \pm\pi/3, \ \pm\pi/2, \ \pm2\pi/3, \ \pm3\pi/4, \ \pm5\pi/6, \ \pm\pi,$$

and check with your diagram that these vectors give the direction of the tangents at the points concerned.
9. Find the tangent vector and the unit tangent vector at a general point for the vector-valued function

$$\mathbf{r} = \cos^2\theta\,\mathbf{i} + \sin^2\theta\,\mathbf{j}.$$

Armed with this information, draw the graph of the function defined by **r**.
10. Suppose $\mathbf{u} = x^2\mathbf{i} + 2x\mathbf{j} + \mathbf{k}$, and $\phi(x) = e^x$.
  (i) Find $d\mathbf{u}/dx$ and $d\phi/dx$.
  (ii) By substituting into Rule 3 for differentiation of vectors, show that

$$\frac{d(\phi\mathbf{u})}{dx} = e^x\{(2x + x^2)\mathbf{i} + (2 + 2x)\mathbf{j} + \mathbf{k}\}.$$

  (iii) Check this by differentiating $\phi\mathbf{u}$ directly.
11. Let $\mathbf{u} = t^2\mathbf{i} + 2t\mathbf{j} - 5\mathbf{k}$, and $\mathbf{v} = \sin t\,\mathbf{i} - \cos t\,\mathbf{j} + \mathbf{k}$. Find:
  (i) $d\mathbf{u}/dt$ and $d\mathbf{v}/dt$,
  (ii) $\mathbf{u}.\mathbf{v}$ and $\mathbf{u} \times \mathbf{v}$, and
  (iii) the derivatives of $\mathbf{u}.\mathbf{v}$ and $\mathbf{u} \times \mathbf{v}$ firstly by using the results of (i) in the RHS of the relevant expressions given in the *rules for differentiating vectors*, and secondly by differentiating directly the results of (ii).

12. Let γ be the curve defined by

$$\mathbf{r} = 2\theta\mathbf{i} + \theta^2\mathbf{j} + \frac{1}{3}\theta^3\mathbf{k}.$$

(i) Find
   (a) the unit tangent vector **t**,
   (b) the unit principal normal **n**,
   (c) the unit binormal **b**,
   (d) the curvature κ and the torsion τ of γ at the point with parameter θ.
(ii) Write down the values of κ and τ when $\theta = 2$.
(iii) Show that the curve lies within the surface $3xz = 2y^2$.
(*Hint*: Remember that $|d\mathbf{r}/ds| = 1$ and that $d\mathbf{r}/ds = (d\mathbf{r}/d\theta)(d\theta/ds)$.)

9 • Non-rectangular Coordinate Systems and Surfaces

Although up to this point we have considered only the usual rectilinear system of axes and coordinates, there are situations where a different coordinate system might be more appropriate, or more convenient. We shall consider one other system in $\mathbb{R}^2$ and two in $\mathbb{R}^3$, but the reader should be aware that there are others.

9.1 Polar coordinates in $\mathbb{R}^2$

For polar coordinates in $\mathbb{R}^2$ we choose an origin O together with a directed *initial line* through O. It is often convenient to think of the initial line as the x-axis of a corresponding rectangular system, but this is not strictly necessary. The polar coordinates of a point P in the plane are (r, θ), where r denotes the distance from the origin, and θ denotes the angle turned through anticlockwise from the initial line. We choose **i** to be a unit vector in the direction of the initial line, and **j** to be a second unit vector defined so that turning from **i** through a right angle anticlockwise would give the direction of **j**.

Here **r** denotes the position vector of a point in the same way as in Chapter 8, and writing $|\mathbf{r}| = r$, we have

$$\mathbf{r} = (r\cos\theta)\mathbf{i} + (r\sin\theta)\mathbf{j}.$$

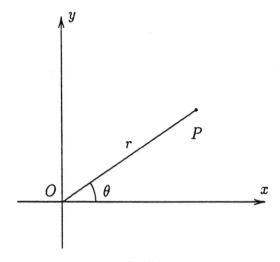

Fig 9.1

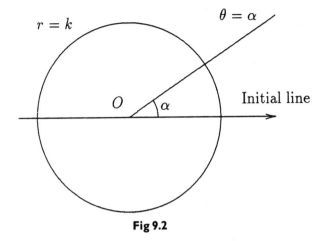

Fig 9.2

With coordinate systems, it is useful to consider what we get if we keep one of the variables constant and let the other(s) range over all possible values. In $\mathbb{R}^2$ with the usual rectangular axes, if c is constant, the equation $x = c$ defines a line, as does $y = c$. (Similarly in $\mathbb{R}^3$, $x = c$ defines a plane.) In $\mathbb{R}^2$, with polar coordinates as described above, $r = c$ defines a circle for $c > 0$, and $\theta = \alpha$ defines a half-line making an angle α with the *initial line* as shown in Fig 9.2.

It is possible to allow r to be negative as well as positive, in which case $r = c$ denotes exactly the same circle as before, but $\theta = \alpha$ denotes the whole line, rather than just a half-line, as negative r would give all the points on the line on the opposite side of the origin from that described above. There are merits in both systems, but since we are using vectors, because we define $r = |\mathbf{r}| \geq 0$, the most appropriate convention in this context is the first, and this is the convention which is used in this book.

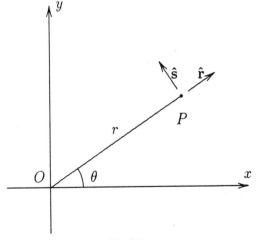

Fig 9.3

Suppose we set the two vectors $\hat{\mathbf{r}}$ and $\hat{\mathbf{s}}$ as the unit vectors in the *radial* and *transverse* directions respectively as shown in Fig. 9.3. Then

$$\hat{\mathbf{r}} = (\cos\theta)\mathbf{i} + (\sin\theta)\mathbf{j}$$

and

$$\hat{\mathbf{s}} = (-\sin\theta)\mathbf{i} + (\cos\theta)\mathbf{j}.$$

EXERCISE I

Suppose that $\mathbf{r}$ and θ are functions of time t. Show that

(i) $\dfrac{d\hat{\mathbf{r}}}{dt} = \dot{\theta}\hat{\mathbf{s}},$ (ii) $\dfrac{d\hat{\mathbf{s}}}{dt} = -\dot{\theta}\hat{\mathbf{r}}.$

Differentiating $\mathbf{r}$ with respect to t component-wise we have

$$\frac{d\mathbf{r}}{dt} = (\dot{r}\cos\theta - r\sin\theta\dot{\theta})\mathbf{i} + (\dot{r}\sin\theta + r\cos\theta\dot{\theta})\mathbf{j} = \dot{r}\hat{\mathbf{r}} + r\dot{\theta}\hat{\mathbf{s}}.$$

Alternatively by direct differentiation of the equation $\mathbf{r} = r\hat{\mathbf{r}}$ we get the last step following from the result of Exercise 1(i).

Differentiating again, and substituting from both results of Exercise 1, we get

$$\ddot{\mathbf{r}} = \frac{d\dot{\mathbf{r}}}{dt} = (\ddot{r} - r\dot{\theta}^2)\hat{\mathbf{r}} + (2\dot{r}\dot{\theta} + r\ddot{\theta})\hat{\mathbf{s}}.$$

9.2 Spherical polar coordinates in $\mathbb{R}^3$

In spherical polars we need to choose an origin O, and a right-handed set of rectangular axes. As in the two-dimensional case, $r = |\mathbf{r}|$ where $\mathbf{r}$ is the position vector of the point P, θ is the angle from the plane $y = 0$ to the plane containing both P and the z-axis, and ϕ is the angle between OP and the z-axis as shown in Fig 9.4. Thus

$$\mathbf{r} = r\cos\theta\sin\phi\mathbf{i} + r\sin\theta\sin\phi\mathbf{j} + r\cos\phi\mathbf{k}.$$

Again we consider what happens when one of the three variables is constant. Suppose k and α are fixed real numbers. Then $r = k$ $(k > 0)$ denotes a sphere of radius k, which is why we refer to coordinates in this system as *spherical polars*. (If $k = 0$ then the 'sphere' is of zero radius, and hence is a single point.) $\theta = \alpha$ is the equation of a half-plane containing the z-axis and a line in the plane $z = 0$ making an angle α with the x-axis as shown, and $\phi = \alpha$ a half-cone with vertex at the origin, and its axis along the z-axis. Typical surfaces of this kind are shown in Fig 9.5.

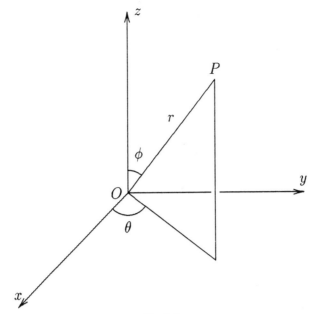

Fig 9.4

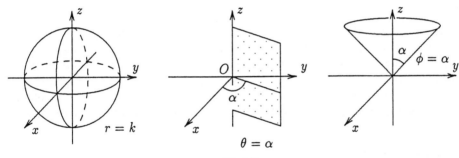

Fig 9.5

EXERCISE 2

Draw diagrams to illustrate the following:

(i) $r = 2$, (ii) $\theta = \pi$, (iii) $\phi = \pi/2$.

EXERCISE 3

Describe geometrically the loci determined by the following pairs of equations:

(i) $r = 2$, $\theta = \pi$, (ii) $r = 2$, $\phi = \pi/2$, (iii) $\theta = \pi$, $\phi = \pi/2$.

9.3 Cylindrical polar coordinates in $\mathbb{R}^3$

Again for cylindrical polars we need to choose an origin O and a right-handed
system of rectangular axes, as in the spherical polar case. Then the coordinates of a

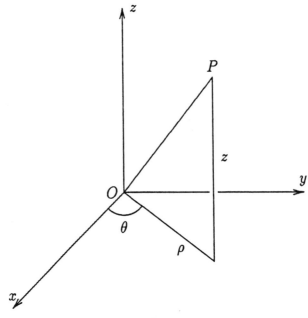

Fig 9.6

general point P in this system are (ρ, θ, z), where ρ is the distance of P from the z-axis, θ is the same angle as in spherical polars, and z gives the usual distance from the plane $z = 0$ as shown in Fig 9.6. Suppose k and α are constant real numbers as before. Then $\rho = k$ ($k > 0$) denotes a cylinder of radius k, which explains the terminology here. (If $k = 0$ we simply get the z-axis.) $\theta = \alpha$ determines a half-plane as in the spherical polar case, and $z = k$ determines a plane perpendicular to the z-axis.

EXERCISE 4

Draw diagrams to illustrate what is defined by each of the following equations:

    (i) $\rho = 2$,   (ii) $\theta = \pi/2$,   (iii) $z = -3$.

EXERCISE 5

Describe geometrically the loci determined by the following pairs of equations:

    (i) $\rho = 2$, $\theta = \pi/2$,   (ii) $\rho = 2$, $z = -3$,   (iii) $\theta = \pi/2$, $z = -3$.

Cylindrical polars combine two-dimensional polar coordinates with a *height z*. If $\mathbf{r}$ is the position vector of P,

$$\mathbf{r} = \rho \cos \theta \, \mathbf{i} + \rho \sin \theta \mathbf{j} + z \mathbf{k}.$$

Our helix in Fig 8.5 is best expressed in terms of cylindrical polar coordinates. Its equations become

$$\rho = a, \qquad z = k\theta, \quad a, k \text{ constant.}$$

The position vector of a general point can therefore be expressed in terms of a single parameter θ and hence the locus is a curve – the helix mentioned above.

9.4 Surfaces

In the same way that a curve is given by the position vector of a general point on the curve in terms of one parameter, so a surface can be given by the position vector of a general point on the surface in terms of two parameters. We have already mentioned two such surfaces, the sphere and the cylinder, when we were discussing spherical and cylindrical polar coordinates. We now look at some examples of surfaces.

Consider, for example, the locus of a point whose position vector is given by

$$\mathbf{r} = x\mathbf{i} + y\mathbf{j} + (x^2 + y^2)\mathbf{k}.$$

The *locus* (literally being translated from Latin as *place*) means the set of all possible positions of that point under the given condition, or, as in this case, satisfying the given equation.

If the coordinates (x, y, z) are those of a point on this locus, then z denotes the *height* of the locus vertically above the point $(x, y, 0)$. As x and y vary, so the point $(x, y, 0)$ can range over the whole of the plane $z = 0$ and as it does so, the point $(x, y, x^2 + y^2)$ ranges over the surface shown in Fig 9.7. If $k > 0$ the plane $z = k$ cuts the surface in a circle whose equations are given by

$$x^2 + y^2 = k, \quad z = k,$$

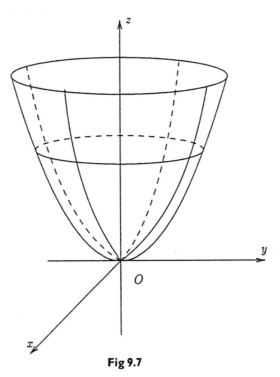

Fig 9.7

whereas if $k = 0$ the intersection is a single point, and if $k < 0$ there is no real intersection. If we look at the intersection with the plane $y = 0$ and $x = 0$ we get the respective parabolas defined by

$$z = x^2, \ y = 0 \qquad \text{and} \qquad z = y^2, \ x = 0.$$

In these directions the intersection is a parabola, and the whole shape is called a circular paraboloid. Typical intersections of types described above are shown in the figure.

In general if we consider the locus of points (x, y) in the plane which are connected by a single equation, the locus is a curve, and if we consider the locus of points (x, y, z) in $\mathbb{R}^3$ which are connected by a single equation, we have a surface. For example,

$$x^2 + y^2 = a^2, \quad a \neq 0$$

defines a circle in $\mathbb{R}^2$, whereas

$$x^2 + y^2 + z^2 = a^2, \quad a \neq 0$$

defines a sphere in $\mathbb{R}^3$. (By a *sphere* we mean the two-dimensional surface. The three-dimensional volume contained within a sphere is called a *ball*.) There are exceptions to this general rule, as we can see if we allow $a = 0$ in the above equations, since in each of the $\mathbb{R}^2$ and $\mathbb{R}^3$ cases the equation will determine a single point, namely the origin.

Using vectors often simplifies equations of well-known surfaces. For example, the equation of the sphere written above could be expressed as $|\mathbf{r}| = a$ or as $\mathbf{r}.\mathbf{r} = a^2$. We have already met the equation of a plane in the form

$$(\mathbf{r} - \mathbf{a}).\mathbf{n} = 0.$$

A sphere of radius a centred at the point B, with position vector $\mathbf{b}$, since $|\mathbf{r} - \mathbf{b}| = a$, has vector equation

$$(\mathbf{r} - \mathbf{b}).(\mathbf{r} - \mathbf{b}) = a^2. \tag{9.2.1}$$

EXERCISE 6

In each of the following cases, describe the surface determined by the given vector equation.

    (i) $(\mathbf{r} - \mathbf{a}).(\mathbf{r} - \mathbf{b}) = 0,$ (ii) $(\mathbf{r} - \mathbf{a}).(\mathbf{r} - \mathbf{b}) = \mathbf{a}.\mathbf{b}.$

9.5 Partial differentiation

A curve has a tangent line at every point, provided the curve is defined by a differentiable function. A surface has a tangent plane at each point, again with the proviso that the surface is defined by a differentiable function. We need to know what this means for surfaces before we can find tangent planes.

Suppose we have a surface S defined by $z = f(x, y)$, where the *height* of the surface is given in terms of the x- and y-coordinates, as described in Section 9.4. Let

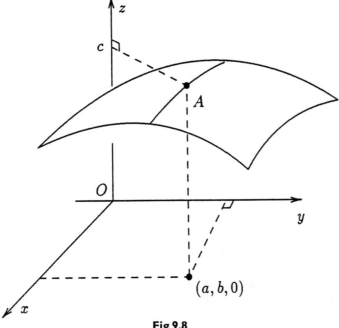

Fig 9.8

$A = (a, b, c)$ be a point on this surface (so that $c = f(a, b)$) and consider the intersection of S with the plane $y = b$. This plane is parallel to the x-axis. Then the plane $y = b$ cuts S in the curve $y = b, z = f(x, b)$, and Fig 9.9 shows the plane $y = b$, and its intersection with S.

The tangent line to this curve of intersection is also shown, and the gradient of this tangent at the point A is

$$\lim_{h \to 0} \frac{f(a + h, b) - f(a, b)}{h}$$

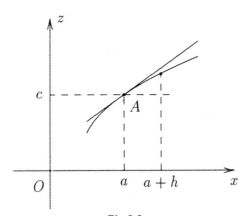

Fig 9.9

if this limit exists. Where this limit does exist we use the notation

$$\lim_{h \to 0} \frac{f(a+h,b) - f(a,b)}{h} = \left.\frac{\partial f}{\partial x}\right|_{(a,b)} = f_x(a,b).$$

We shall use the latter notation because it is more convenient to use with vectors. In this book we shall consider only functions which are differentiable, but the reader should be aware that there are functions in two variables which do not behave *nicely* just as there are with functions of one variable. (One example is $y = 1/x$ which shoots off to infinity as x approaches 0, and another is $y = |x|$ which is not differentiable at $x = 0$, as it is not *smooth* there.)

Similarly if we consider the intersection of S with the plane $x = a$, the gradient to the curve of intersection at the point A is

$$\lim_{h \to 0} \frac{f(a,b+k) - f(a,b)}{k} = f_y(a,b).$$

If we are finding the partial derivatives at a general point we write f_x rather than $f_x(x,y)$, as this makes the notation less unwieldy. In practice, to find f_x we simply differentiate $f(x,y)$ with respect to x treating y as a constant (which it *is* within the plane $y = b$), and to find f_y we differentiate f with respect to y, treating x as a constant.

● *Example 1*

Suppose $f(x,y) = x^2y^3 + 2x - 3y$. Find $f_x(1,2)$ and $f_y(2,-3)$.

SOLUTION
$f_x = 2xy^3 + 2$, treating the y^3 and the $-3y$ as constants. We now substitute the values for x and y, so that

$$f_x(1,2) = 2.1.2^3 + 2 = 18.$$

$f_y = 3x^2y^2 - 3$, again treating the x^2 and the $2x$ as constants, and so

$$f_y(2,-3) = 3.2^2.(-3)^2 - 3 = 105.$$

EXERCISE 7

Given that $f(x,y) = x^2y + 2xy^3 - xy$, find f_x and f_y and hence evaluate $f_x(2,-2)$ and $f_y(3,-1)$.

9.6 Tangent planes

Because the lines in the planes $x = a$ and $y = b$ which are tangent to the intersections of these respective planes with S must of necessity be tangent to the surface S, they must both lie in the tangent plane to S at A.

Suppose the tangent plane to S at A is π whose equation is

$$px + qy + rz = s;$$

then since A lies on π

$$pa + qb + rc = s$$

where $c = f(a, b)$. We can now write the equation of π as

$$p(x - a) + q(y - b) + r(z - c) = 0.$$

This intersects the plane $y = b$ in the line

$$y = b, \qquad p(x - a) + r(z - c) = 0.$$

But this is the line $y = b$, $z - c = f_x(a, b)(x - a)$, so $-1 : f_x(a, b) = r : p$. Similarly π intersects the plane $x = a$ in the line

$$x = a, \qquad q(y - b) + r(z - c) = 0$$

which is the line $x = a$, $z - c = f_y(a, b)(y - b)$, and so $-1 : f_y(a, b) = r : q$. This means that the equation of the tangent plane to S at A is

$$f_x(a, b)(x - a) + f_y(a, b)(y - b) - (z - c) = 0. \tag{9.6.1}$$

This also means that the vector normal to the surface at A is parallel to the vector

$$\mathbf{n} = f_x(a, b)\mathbf{i} + f_y(a, b)\mathbf{j} - \mathbf{k}.$$

Suppose the position vector of A is $\mathbf{a}$, and that of a general point on the surface is $\mathbf{r}$; then we can write the vector equation of the tangent plane as

$$(\mathbf{r} - \mathbf{a}).\mathbf{n}.$$

We do not fall into the danger of having the coefficient of z equal to zero, as this would mean that the surface would have a vertical tangent plane at the point A, which we have avoided by assuming that f is a differentiable function.

Example 2

Find the equation of the tangent plane to the surface defined by

$$z = x^2 + 2y^2 - 6xy + 4$$

at the point on the surface where $x = 0, y = 1$.

SOLUTION
If we set $f(x, y) = x^2 + 2y^2 - 6xy + 4$, then $f_x = 2x - 6y$ and $f_y = 4y - 6x$, so that

$$f_x(0, 1) = 0 - 6 = -6, \quad f_y(0, 1) = 4 - 0 = 4,$$

and $f(0, 1) = 6$.
    Hence from equation (9.6.1), the equation of the tangent plane at the given point is

$$-6(x - 0) + 4(y - 1) - (z - 6) = 0$$

which reduces to

$$6x - 4y + z + 2 = 0.$$

EXERCISE 8

Find the equation of the tangent plane to the surface defined by

$$z = x^2 y + 2y^2 - 3xy$$

at the point on the surface where $x = 1, y = 2$.

We can deal with surfaces defined by implicit functions, and these may involve some tangent planes parallel to any of the axes of symmetry. Suppose the surface is defined by an equation of the form

$$F(x, y, z) = 0. \tag{9.6.2}$$

We can choose any two of the variables x, y and z independently, and then because of the relation defined by equation (9.6.2) the third variable is dependent upon these two. Suppose we choose x and y as the independent variables. Then z is dependent upon both x and y. Thus, if we differentiate equation (9.6.2) partially with respect to x, we get

$$F_x + F_z \frac{\partial z}{\partial x} = 0$$

and similarly, differentiating partially with respect to y,

$$F_y + F_z \frac{\partial z}{\partial y} = 0$$

and, provided $F_z \neq 0$,

$$\frac{\partial z}{\partial x} = \frac{-F_x}{F_z} \quad \text{and} \quad \frac{\partial z}{\partial y} = \frac{-F_y}{F_z}. \tag{9.6.3}$$

Substituting from equations (9.6.3) into equation (9.6.1) and multiplying through by F_z we can write the equation of the tangent plane to S at A as

$$F_x(a, b, c)(x - a) + F_y(a, b, c)(y - b) + F_z(a, b, c)(z - c) = 0. \tag{9.6.4}$$

Alternatively, if we define ∇F to be the vector $F_x \mathbf{i} + F_y \mathbf{j} + F_z \mathbf{k}$, together with the usual convention for position vectors of A and the general point on S, we have

$$\nabla F(\mathbf{a}) \cdot (\mathbf{r} - \mathbf{a}) = 0. \tag{9.6.5}$$

The vector $\nabla F(\mathbf{a})$ is *normal* to the surface S at A, since it is normal (that is, perpendicular) to the tangent plane at A. Even if $F_z = 0$, equation (9.6.4) can still be used as, on a differentiable surface, we cannot have all three partial derivatives zero at the same time. If $F_x \neq 0$, for example, we could simply exchange the rôles of x and z, taking y and z to be the independent variables, and x to be dependent upon both of these. Then we should end up with equation (9.6.4) just the same.

Example 3

Suppose a surface S is defined by

$$F(x, y, z) = x^2 + y^2 + z^2 - 9. \tag{9.6.6}$$

Find the equation of the tangent plane to S at the point A whose coordinates are $(1, 2, 2)$.

SOLUTION

Firstly it is a good idea to check that the point really does lie on S.

$$1^2 + 2^2 + 2^2 - 9 = 0$$

which verifies that A lies on S. Then

$$F_x = 2x, \; F_y = 2y, \; F_z = 2z$$

and substituting into equation (9.6.4), using the values $x = 1, y = 2, z = 2$, we get

$$2(x - 1) + 4(y - 2) + 4(z - 2) = 0$$

which, tidied up, becomes

$$x + 2y + 2z = 9.$$

Now, in this example, equation $F(x, y, z) = 0$ defines a sphere with centre at the origin, and radius 3.

As a check that we have arrived at the correct answer, we can note that a vector normal to the tangent plane is $\mathbf{i} + 2\mathbf{j} + 2\mathbf{k}$ which is parallel to the position vector of A. But since the centre of the sphere lies at the origin, and the position vector of A thus lies along a radius of the sphere, it must be perpendicular to the tangent plane at the point of contact, namely A, and so this consolidates our answer.

EXERCISE 9

Find the equation of the tangent plane at the point A to the surface S which is defined by $F(x, y, z) = 0$ when

$$F(x, y, z) = 2x^2 - 3y^2 + xy + 2z$$

and A has coordinates $(1, -2, 6)$. Write down a vector which is normal to the surface S at A.

The methods described above will be valid for any surface S which is defined by a function which has continuous first partial derivatives, but it is not within the scope of this book to approach, for example, surfaces containing any singular points or lines such as cusps or even more exotic entities.

9.7 Gradient, divergence and curl

We have referred to the mechanical implications of properties of vectors many times, and to go further into this subject we would need to look at three different types of derivative *grad*, *div* and *curl*. We have actually already met *grad* in Section 9.6 where we used it in the vector equation of a tangent plane to a surface at a given point. The three derivatives are very much related and depend upon what is called the *vector differential operator* ∇, which we call *del* (although some texts refer to this as *nabla*), and which can be defined as

$$\nabla = \frac{\partial}{\partial x}\mathbf{i} + \frac{\partial}{\partial y}\mathbf{j} + \frac{\partial}{\partial z}\mathbf{k}.$$

This can *act on* scalar or vector functions giving the three types of derivative as follows.

Gradient

Suppose $\phi : A \rightarrow \mathbb{R}$ is a differentiable function, where A is a region of $\mathbb{R}^3$ (so that $A \subset \mathbb{R}^3$ and at each point (x, y, z) of A the value of the function $\phi(x, y, z)$ is a real number); then the *gradient* of ϕ, denoted by *grad* ϕ or $\nabla\phi$, is defined by

$$\nabla\phi = \frac{\partial\phi}{\partial x}\mathbf{i} + \frac{\partial\phi}{\partial y}\mathbf{j} + \frac{\partial\phi}{\partial z}\mathbf{k}.$$

$\nabla\phi$ defines what is called a *vector field* on A, that is at every point in A a vector $\nabla\phi$ is defined, and small variations in position correspond to small variations in the corresponding vector. We see an example of a vector field when we watch the television weather forecaster's wind-chart. The arrows show the direction of the wind, and the size of the arrows shows the magnitude of the windspeed. If we think of this idea extended to every point of a three-dimensional region within $\mathbb{R}^3$ instead of only those points on the two-dimensional map of the British Isles, then we have some concept of a three-dimensional vector field.

If $\mathbf{u}$ is a unit vector, then $\nabla\phi.\mathbf{u}$, the component of $\nabla\phi$ in the direction of $\mathbf{u}$, is called the *directional derivative* of ϕ in the direction of $\mathbf{u}$. It could tell us, for example, how density is increasing in a particular direction.

Divergence

Suppose $\mathbf{V}$ is a differentiable vector-valued function defined on a region A of $\mathbb{R}^3$, so that

$$\mathbf{V}(x, y, z) = \phi_1\mathbf{i} + \phi_2\mathbf{j} + \phi_3\mathbf{k}$$

where ϕ_1, ϕ_2 and ϕ_3 are differentiable real-valued functions on A. This means that $\phi_i(x, y, z)$, for $i = 1, 2, 3$, is a real number for all (x, y, z) in A. The *divergence* of $\mathbf{V}$, denoted by *div* $\mathbf{V}$ or $\nabla.\mathbf{V}$, is given by

$$\nabla.\mathbf{V} = \left(\frac{\partial}{\partial x}\mathbf{i} + \frac{\partial}{\partial y}\mathbf{j} + \frac{\partial}{\partial z}\mathbf{k}\right).(\phi_1\mathbf{i} + \phi_2\mathbf{j} + \phi_3\mathbf{k}) = \frac{\partial\phi_1}{\partial x} + \frac{\partial\phi_2}{\partial y} + \frac{\partial\phi_3}{\partial z}.$$

Note that this behaves like a scalar product, which is why the *dot* notation is used.

Curl

Suppose **V** is a differentiable vector function as above; then the *curl* of **V**, denoted by *curl* **V** or $\nabla \times \mathbf{V}$, is defined by

$$\nabla \times \mathbf{V} = \left(\frac{\partial}{\partial x}\mathbf{i} + \frac{\partial}{\partial y}\mathbf{j} + \frac{\partial}{\partial z}\mathbf{k} \right) \times (\phi_1\mathbf{i} + \phi_2\mathbf{j} + \phi_3\mathbf{k})$$

$$= \left(\frac{\partial \phi_3}{\partial y} - \frac{\partial \phi_2}{\partial z} \right)\mathbf{i} + \left(\frac{\partial \phi_1}{\partial z} - \frac{\partial \phi_3}{\partial x} \right)\mathbf{j} + \left(\frac{\partial \phi_2}{\partial x} - \frac{\partial \phi_1}{\partial y} \right)\mathbf{k}.$$

This is sometimes referred to as the *rotation* of **V**, and denoted by *rot* **V**.

9.8 Further study

Vectors are useful in most branches of mathematics. Even within statistics and operational research they appear from time to time. As part of the basic toolset in mechanics they can be used for solving problems involving forces in three dimensions, motion of a particle or a body, or even planetary motion. In spherical geometry the whole idea of isometries is made easier by regarding all the points on the sphere as being defined by unit position vectors.

The derivatives in the previous section have been included to show how central the ideas of scalar product and vector product are in the study of vector analysis at all levels. They would be needed in vector integration, particularly in the applications of important theorems such as the Divergence Theorem, and the theorems of Stokes and Green. Further study might also involve tensor calculus where ideas analogous to the vector product in higher dimensions are involved.

There is a great deal of calculus involved in the further study of geometrical vectors. Nevertheless once all the concepts included here have been digested and understood, the journey through vector analysis to differential geometry is an intriguing and rewarding one, and I would highly recommend it.

Summary

1. The point with polar coordinates (r, θ) in $\mathbb{R}^2$ has position vector

    $(r\cos\theta)\mathbf{i} + (r\sin\theta)\mathbf{j}.$

2. The point with spherical polar coordinates (r, θ, ϕ) in $\mathbb{R}^3$ has position vector

    $(r\cos\theta\sin\phi)\mathbf{i} + (r\sin\theta\sin\phi)\mathbf{j} + (r\cos\phi)\mathbf{k}.$

3. The point with cylindrical polar coordinates (ρ, θ, z) in $\mathbb{R}^3$ has position vector

    $(\rho\cos\theta)\mathbf{i} + (\rho\sin\theta)\mathbf{j} + z\mathbf{k}.$

4. The tangent plane to a surface whose equation is $z = f(x, y)$ at a point A (with position vector **a**) on the surface is

    $f_x(a, b)(x - a) + f_y(a, b)(y - b) - (z - c) = 0$

    where $c = f(a, b)$ and $\mathbf{a} = a\mathbf{i} + b\mathbf{j} + c\mathbf{k}.$

5. The tangent plane to a surface whose equation is $F(x, y, z) = 0$ at a point A (with position vector $\mathbf{a}$) on the surface is

$$F_x(a, b, c)(x - a) + F_y(a, b, c)(y - b) + F_z(a, b, c)(z - c) = 0$$

where $F(a, b, c) = 0$ and $\mathbf{a} = a\mathbf{i} + b\mathbf{j} + c\mathbf{k}$.

6. The normal vector to a surface whose equation is $F(x, y, z) = 0$ at a point A (with position vector $\mathbf{a}$) on the surface is $\nabla F(\mathbf{a})$, where

$$\nabla F(\mathbf{a}) = F_x(a, b, c)\mathbf{i} + F_y(a, b, c)\mathbf{j} + F_z(a, b, c)\mathbf{k}.$$

The vector equation of the tangent plane to the surface at A is

$$\nabla F(\mathbf{a}) \cdot (\mathbf{r} - \mathbf{a}) = 0.$$

7. Suppose $\phi : A \rightarrow \mathbb{R}$ is a differentiable scalar function, and $\mathbf{V}$ is a differentiable vector function defined on a region A of $\mathbb{R}^3$, where

$$\mathbf{V}(x, y, z) = \phi_1\mathbf{i} + \phi_2\mathbf{j} + \phi_3\mathbf{k};$$

then *grad*, *div* and *curl* are defined respectively by

$$grad\ \phi = \nabla\phi = \frac{\partial\phi}{\partial x}\mathbf{i} + \frac{\partial\phi}{\partial y}\mathbf{j} + \frac{\partial\phi}{\partial z}\mathbf{k},$$

$$div\ \mathbf{V} = \nabla . \mathbf{V} = \frac{\partial\phi_1}{\partial x} + \frac{\partial\phi_2}{\partial y} + \frac{\partial\phi_3}{\partial z},$$

$$curl\ \mathbf{V} = \nabla \times \mathbf{V} = \left(\frac{\partial\phi_3}{\partial y} - \frac{\partial\phi_2}{\partial z}\right)\mathbf{i} + \left(\frac{\partial\phi_1}{\partial z} - \frac{\partial\phi_3}{\partial x}\right)\mathbf{j} + \left(\frac{\partial\phi_2}{\partial x} - \frac{\partial\phi_1}{\partial y}\right)\mathbf{k}.$$

FURTHER EXERCISES

10. Describe geometrically the figures defined by the following *pairs* of equations in spherical polar coordinates, and draw a diagram to illustrate each of these cases.

(i) $r = 2,\ \theta = \pi/4$, (ii) $r = 3,\ \phi = \pi/4$, (iii) $\phi = \pi/4,\ \theta = \pi/4$.

11. Describe geometrically the figures defined by the following pairs of equations in cylindrical polar coordinates, and draw a diagram to illustrate each of these cases.

(i) $\rho = 1,\ \theta = \pi/6$, (ii) $\rho = 2,\ z = -1$, (iii) $\theta = \pi/3,\ z = 2$.

12. Given that $f(x, y) = x^2y - 2xy + y^3$, find the tangent plane to the surface defined by the equation $z = f(x, y)$ at the point where $x = 1, y = -2$.

13. Given that $F(x, y, z) = x^2yz + 2xy^3 + 4y^2z^2 - 1$, find
    (i) F_x, F_y and F_z,
    (ii) the tangent to the surface defined by $F(x, y, z) = 0$ at the point $(1, -1, 1)$, and
    (iii) a unit vector normal to the surface $F(x, y, z) = 0$ at the point $(1, -1, 1)$.

14. If $\phi : \mathbb{R}^3 \rightarrow \mathbb{R}$ is defined by

$$\phi(x, y, x) = e^x(y^2 - yz),$$

find $\nabla\phi$, and evaluate this at the point $(1, 2, -3)$.

15. If

$$\mathbf{V} = (x^2 y^2 + \cos z)\mathbf{i} + (2xy^3 + \sin z)\mathbf{j} - (3x^3 y)\mathbf{k}$$

find (i) $\nabla \cdot \mathbf{V}$ and (ii) $\nabla \times \mathbf{V}$.

Answers to Exercises

Chapter 1

1. $\mathbf{b} = 2\mathbf{a}$, $\mathbf{c} = 3\mathbf{a}$, $\mathbf{d} = -\mathbf{a}$, $\mathbf{e} = \frac{1}{2}\mathbf{a}$, $\mathbf{f} = -\mathbf{a}$, $\mathbf{g} = 3\mathbf{a}$, $\mathbf{h} = \frac{3}{2}\mathbf{a}$.

2. (ii) and (iv) are RH, others are LH.

3. $\mathbf{a} = 2\mathbf{j}$, $\mathbf{b} = 2\mathbf{k}$, $\mathbf{c} = 2\mathbf{k} - \mathbf{i} - \mathbf{j}$, $\mathbf{d} = \mathbf{k} - \mathbf{i}$, $\mathbf{e} = 2\mathbf{j} - 2\mathbf{k}$, $\mathbf{f} = 2\mathbf{i} + \mathbf{j} - 2\mathbf{k}$, $\mathbf{g} = 3\mathbf{i} - \mathbf{j} - \mathbf{k}$, $\mathbf{h} = \mathbf{i} - \mathbf{j}$.

4. $\begin{pmatrix} 0 \\ 2 \\ 0 \end{pmatrix}$, $\begin{pmatrix} 0 \\ 0 \\ 2 \end{pmatrix}$, $\begin{pmatrix} -1 \\ -1 \\ 2 \end{pmatrix}$, $\begin{pmatrix} -1 \\ 0 \\ 1 \end{pmatrix}$, $\begin{pmatrix} 0 \\ 2 \\ -2 \end{pmatrix}$, $\begin{pmatrix} 2 \\ 1 \\ -2 \end{pmatrix}$, $\begin{pmatrix} 3 \\ -1 \\ -1 \end{pmatrix}$, $\begin{pmatrix} 1 \\ -1 \\ 0 \end{pmatrix}$.

5. (i) $\sqrt{3}$, (ii) 3, (iii) 7, (iv) $5\sqrt{2}$.

6. (i) $\begin{pmatrix} 3 \\ 0 \\ 4 \end{pmatrix}$, (ii) $\begin{pmatrix} 8 \\ -2 \\ 9 \end{pmatrix}$, (iii) $\begin{pmatrix} -6 \\ 18 \\ 7 \end{pmatrix}$, (iv) $\begin{pmatrix} 5/3 \\ -2/3 \\ 5/3 \end{pmatrix}$.

9. $\mathbf{c}$ and $\mathbf{f}$ are parallel; $\mathbf{e}$ and $\mathbf{g}$ are parallel.

10. $|\mathbf{c}| = \sqrt{14}$, $|\mathbf{d}| = \sqrt{14}$, $|\mathbf{e}| = \sqrt{3/2}$, $|\mathbf{f}| = \sqrt{14}/6$, $|\mathbf{g}| = \sqrt{6}/4$.

11. $\mathbf{c} = 8\mathbf{i} + \mathbf{j} - \mathbf{k}$, $\mathbf{d} = 4\mathbf{i} + 3\mathbf{j} - 5\mathbf{k}$, $\mathbf{e} = 20\mathbf{i} + 5\mathbf{j} - 7\mathbf{k}$, $\mathbf{f} = 22\mathbf{i} + 4\mathbf{j} - 5\mathbf{k}$, $\mathbf{g} = 10\mathbf{j} - 18\mathbf{k}$.

12. $\hat{\mathbf{a}} = \dfrac{1}{3}(2\mathbf{i} - \mathbf{j} + 2\mathbf{k})$, $\hat{\mathbf{b}} = \dfrac{1}{7}(6\mathbf{i} + 2\mathbf{j} - 3\mathbf{k})$, $\hat{\mathbf{c}} = \dfrac{1}{\sqrt{66}}(8\mathbf{i} + \mathbf{j} - \mathbf{k})$,

$\hat{\mathbf{d}} = \dfrac{1}{5\sqrt{2}}(4\mathbf{i} + 3\mathbf{j} - 5\mathbf{k})$, $\hat{\mathbf{e}} = \dfrac{1}{\sqrt{474}}(20\mathbf{i} + 5\mathbf{j} - 7\mathbf{k})$, $\hat{\mathbf{f}} = \dfrac{1}{5\sqrt{21}}(22\mathbf{i} + 4\mathbf{j} - 5\mathbf{k})$,

$\hat{\mathbf{g}} = \dfrac{1}{\sqrt{106}}(5\mathbf{j} - 9\mathbf{k})$.

Chapter 2

1. $\mathbf{r} = \begin{pmatrix} 1 \\ 2 \\ -3 \end{pmatrix} + \lambda \begin{pmatrix} 2 \\ 1 \\ -5 \end{pmatrix}$, $\dfrac{x-1}{2} = \dfrac{y-2}{1} = \dfrac{z+3}{-5}$.

3. (i) $\begin{pmatrix} 2 \\ 0 \\ -1 \end{pmatrix}$, (ii) $6 : -1$.

4. $V = (4, 7, 6)$; D divides AV in the ratio $1 : 3$; E divides BV in the ratio $1 : 2$; F divides CV in the ratio $2 : 3$.

5. (i) $\mathbf{r} = \begin{pmatrix} 2 \\ 1 \\ 2 \end{pmatrix} + \lambda \begin{pmatrix} 4 \\ -2 \\ 3 \end{pmatrix}$, $\dfrac{x-2}{4} = \dfrac{y-1}{-2} = \dfrac{z-2}{3}$.

(ii) $\mathbf{r} = (1-\lambda)\begin{pmatrix} 3 \\ -1 \\ 2 \end{pmatrix} + \lambda \begin{pmatrix} 1 \\ 2 \\ 4 \end{pmatrix}$, $\dfrac{x-3}{-2} = \dfrac{y+1}{3} = \dfrac{z-2}{2}$.

6. On AB are $(2,0,1)$, $(0,-2,-1)$; on CD $(2,-2,4)$; the rest are on neither.

7. (i) Lines meet at $(1,-1,1)$, (ii) lines meet at $(38/9, 13/9, 56/9)$, (iii) lines do not meet.

9. (i) $\mathbf{v} = \begin{pmatrix} 3/2 \\ 3/2 \\ 6 \end{pmatrix}$; $AV = 9\sqrt{2}/2$.

(ii) $\mathbf{r} = \begin{pmatrix} \lambda \\ \lambda \\ 4\lambda \end{pmatrix}$; $\mathbf{r} = \begin{pmatrix} 3-\lambda \\ \lambda \\ 4\lambda \end{pmatrix}$; $\mathbf{r} = \begin{pmatrix} 3-\lambda \\ 3-\lambda \\ 4\lambda \end{pmatrix}$; $\mathbf{r} = \begin{pmatrix} \lambda \\ 3-\lambda \\ 4\lambda \end{pmatrix}$.

(iii) 21.9 m; 22 m.

Chapter 3

1. (i) 3, (ii) 2.
2. Orthogonal pairs are $\mathbf{a}$ and $\mathbf{c}$, $\mathbf{a}$ and $\mathbf{d}$, $\mathbf{b}$ and $\mathbf{d}$, $\mathbf{b}$ and $\mathbf{e}$, $\mathbf{c}$ and $\mathbf{e}$.
3. $3/\sqrt{14}$.
4. (i) $74°$, (ii) $112°$.
5. $AB = \sqrt{35}$, $AC = \sqrt{46}$, $BC = \sqrt{33}$, $\angle A = 53.3°$, $\angle B = 71.1°$, $\angle C = 55.6°$.

6. $\left(\mathbf{r} - \begin{pmatrix} 4 \\ -1 \\ 2 \end{pmatrix}\right) \cdot \begin{pmatrix} 1 \\ 3 \\ -2 \end{pmatrix} = 0$ or $x + 3y - 2z + 3 = 0$.

7. (i) $\dfrac{x+1}{-5} = \dfrac{y - \frac{1}{2}}{1} = \dfrac{z-2}{3}$,

(ii) planes parallel – no intersection,

(iii) $\dfrac{x}{-1} = \dfrac{y + 11/5}{2} = \dfrac{z + 4/5}{2}$,

(iv) planes coincide.

8. (i) 5, 2, 7; (ii) $16°$, $77°$, $63°$.
9. $PQ = RQ = 7$; hence $\triangle PQR$ is isosceles ($PR = \sqrt{34}$, so it is not equilateral). $\angle P = 65.4°$, $\angle Q = 49.2°$, $\angle R = 65.4°$.
10. (i) $12/\sqrt{30}$, (ii) $12/\sqrt{29}$, (iii) 12, (iv) 12, (v) both (iii) and (iv) evaluate $\mathbf{a.b}$, and so are equal.
11. $9/\sqrt{83}$, $-1/\sqrt{83}$, $1/\sqrt{83}$; $-1/\sqrt{35}$, $-3/\sqrt{35}$, $5/\sqrt{35}$.
12. (i) $3x - y + 3z = 20$; (ii) $34x - y + 9z = 81$.

13. $\dfrac{x}{-5} = \dfrac{y - 12/5}{13} = \dfrac{z - 11/5}{14}.$

14. (i) Single point $(1,2,3)$, (ii) $\emptyset$ – second plane parallel to first, (iii) line
$$\mathbf{r} = \begin{pmatrix} 4/3 \\ 1/3 \\ 0 \end{pmatrix} + \lambda \begin{pmatrix} -1 \\ 2 \\ 3 \end{pmatrix},\ (iv)\ \emptyset - \text{planes intersect in pairs in three parallel lines.}$$

15. (i) Plane through O, perpendicular to OA, (ii) plane through A, perpendicular to OA, (iii) sphere, with AB as diameter, (iv) sphere, centre at origin, radius OA (passes through A), (v) sphere, centre at A, radius OA (passes through O).

Chapter 4

1. (i) $4\mathbf{i} - 7\mathbf{j} - 13\mathbf{k}$, (ii) $-2\mathbf{i} + \mathbf{j} + \mathbf{k}$, (iii) $-2\mathbf{i} - 6\mathbf{j} - 10\mathbf{k}$.

2. $\left(\mathbf{r} - \begin{pmatrix} 1 \\ -2 \\ 2 \end{pmatrix} \right) \cdot \begin{pmatrix} 7 \\ 4 \\ -9 \end{pmatrix} = 0$ or $7x + 4y - 9z + 19 = 0$.

3. $\sqrt{2}.$

4. (i) $9/\sqrt{10}$, (ii) $\sqrt{10}$.

5. (i) $\mathbf{r} = \begin{pmatrix} 2 \\ 1 \\ 0 \end{pmatrix} + \lambda \begin{pmatrix} -1 \\ 8 \\ 5 \end{pmatrix}$; (ii) $\mathbf{r} = \begin{pmatrix} 3 \\ 2 \\ 1 \end{pmatrix} + \lambda \begin{pmatrix} 2 \\ -1 \\ 0 \end{pmatrix}.$

7. (i) $\lambda = 27$, (ii) using (4.7.1) $\lambda = (\mathbf{a} \times \mathbf{b}) . \mathbf{c}$.

8. (i) $-7\mathbf{i} + 2\mathbf{j} + 10\mathbf{k}$, (ii) $-6\mathbf{i} - 10\mathbf{j} - 7\mathbf{k}$.

9. 14 (cubic units).

10. Distance of P from l is $2\sqrt{2}/\sqrt{3}$, $\mathbf{q} = \frac{7}{3}\mathbf{i} - \frac{4}{3}\mathbf{j} - \frac{1}{3}\mathbf{k}$.

11. (i) $13/\sqrt{326}$, (ii) $\sqrt{2}$.

12. $\mathbf{r} = \begin{pmatrix} 1 \\ 1 \\ 1 \end{pmatrix} + \lambda \begin{pmatrix} 10 \\ -3 \\ -11 \end{pmatrix}$ or $\dfrac{x - 1}{10} = \dfrac{y - 1}{-3} = \dfrac{z - 1}{-11}.$

Note: Fixed points may differ, but direction vector for the line should be the same in all cases.

13. (i) $\mathbf{r} = \begin{pmatrix} 2 \\ 0 \\ 0 \end{pmatrix} + \lambda \begin{pmatrix} 1 \\ -2 \\ -1 \end{pmatrix}$; $\mathbf{r} = \begin{pmatrix} 1 \\ 1 \\ 0 \end{pmatrix} + \lambda \begin{pmatrix} 0 \\ 1 \\ 1 \end{pmatrix}$; $\mathbf{r} = \begin{pmatrix} 1 \\ 2 \\ 1 \end{pmatrix} + \lambda \begin{pmatrix} 1 \\ 10 \\ 8 \end{pmatrix}.$

The point $(1, 2, 1)$ lies on all three lines. Perpendiculars to three planes are not coplanar.

(ii) $\mathbf{r} = \begin{pmatrix} 2 \\ 2 \\ 0 \end{pmatrix} + \lambda \begin{pmatrix} 4 \\ -1 \\ -3 \end{pmatrix}$; $\mathbf{r} = \begin{pmatrix} 1/2 \\ -1 \\ 0 \end{pmatrix} + \lambda \begin{pmatrix} 4 \\ -1 \\ -3 \end{pmatrix}$; $\mathbf{r} = \begin{pmatrix} -1 \\ 5 \\ 0 \end{pmatrix} + \lambda \begin{pmatrix} 4 \\ -1 \\ -3 \end{pmatrix}.$

The three lines are parallel, but not coincident (they cut the plane $z = 0$ in three different points). Hence there is no point on all three planes.

14. (i) -14, (ii) $-76\mathbf{i} - 32\mathbf{j} + 40\mathbf{k}$, (iii) $-42\mathbf{i} - 28\mathbf{j} + 56\mathbf{k} = -14\mathbf{a}$.

15. (i) 1, (ii) 0.

16. Either $(-2, 2, 1)$, $(3, 3, 0)$, $(0, 3, 3)$, $(-1, 4, -1)$, $(1, 5, 1)$
    or $(2, -2, -1)$, $(3, 3, 0)$, $(4, -1, 1)$, $(3, 0, -3)$, $(5, -1, 1)$.

    Planes are either

    $$
    \begin{aligned}
    2x + \ y + 2z &= 0 & 2x + \ y + 2z &= 9 \\
    x + 2y - 2z &= 0 & x + 2y - 2z &= 9 \\
    -2x + 2y + \ z &= 0 & -2x + 2y + \ z &= 9
    \end{aligned}
    $$

    or

    $$
    \begin{aligned}
    2x + \ y + 2z &= 0 & 2x + \ y + 2z &= 9 \\
    x + 2y - 2z &= 0 & x + 2y - 2z &= 9 \\
    -2x + 2y + \ z &= 0 & 2x - 2y - \ z &= 9.
    \end{aligned}
    $$

Chapter 5

3. (i) Single point at the origin, subspace; (ii) sphere, centre at origin, radius 1, not a subspace (not closed under scalar multiplication); (iii) line through the origin and $(1,1,1)$, subspace.

4. (i) Not a spanning set, (ii) a spanning set.

5. (i) Not linearly independent, not spanning, (ii) linearly independent and spanning, (iii) spanning but not linearly independent, (iv) neither spanning nor linearly independent.

6. (i) Not a basis (not linearly independent), (ii) basis.

7. $\begin{pmatrix} 2 \\ 1 \\ 4 \end{pmatrix} = \dfrac{5}{3} \begin{pmatrix} 1 \\ -1 \\ 1 \end{pmatrix} + (-1) \begin{pmatrix} 1 \\ 0 \\ -1 \end{pmatrix} + \dfrac{4}{3} \begin{pmatrix} 1 \\ 2 \\ 1 \end{pmatrix}.$

9. $\mathbf{u}_1 = \begin{pmatrix} 1 \\ 1 \\ 1 \\ 0 \end{pmatrix}, \qquad \mathbf{u}_2 = \begin{pmatrix} -1 \\ 0 \\ 1 \\ 0 \end{pmatrix}, \qquad \mathbf{u}_3 = \begin{pmatrix} 1 \\ -2 \\ 1 \\ 4 \end{pmatrix}.$

10. $\hat{\mathbf{u}}_1 = \begin{pmatrix} 1/\sqrt{3} \\ 1/\sqrt{3} \\ 1/\sqrt{3} \\ 0 \end{pmatrix}, \qquad \hat{\mathbf{u}}_2 = \begin{pmatrix} -1/\sqrt{2} \\ 0 \\ 1/\sqrt{2} \\ 0 \end{pmatrix}, \qquad \mathbf{u}_3 = \begin{pmatrix} 1/\sqrt{22} \\ -2/\sqrt{22} \\ 1/\sqrt{22} \\ 4/\sqrt{22} \end{pmatrix}.$

11. (i) $3\sqrt{2}$, 3, 2, $2\sqrt{3}$; (ii) 0, 8, 0, 1, -3, 0; (iii) $\mathbf{a}$ and $\mathbf{b}$, $\mathbf{a}$ and $\mathbf{d}$, $\mathbf{c}$ and $\mathbf{d}$.

12. (i) S is a plane through the origin, subspace, (ii) T is a line through the origin (namely the x-axis), subspace, (iii) U is the union of the three axes (like a 3-D cross), not a subspace.

13. (i) Yes, $\mathbf{a} = 2\mathbf{u} + 2\mathbf{v}$, (ii) no, (iii) yes, $\mathbf{c} = \dfrac{7}{5}\mathbf{u} - \dfrac{1}{5}\mathbf{v}$.

14. (i) Yes, two linearly independent vectors, (ii) no, linearly dependent, (iii) no, not a spanning set, (iv) neither spanning nor linearly independent.

15. $\mathbf{x} = \dfrac{3}{2}\mathbf{u}_1 + \dfrac{(-1)}{3}\mathbf{u}_2 + \dfrac{(-1)}{6}\mathbf{u}_3.$

16. (i) $1/\sqrt{6}, -2/\sqrt{6}, 1/\sqrt{6}$; (ii) $\mathbf{p} \times \mathbf{q} = \begin{pmatrix} 1 \\ -2 \\ 1 \end{pmatrix} = \sqrt{6}\,\mathbf{x}$; $\mathbf{p} \times \mathbf{q}$ and $\mathbf{x}$ are parallel

because both are perpendicular to both $\mathbf{p}$ and $\mathbf{q}$.

(iii) $\mathbf{u}_1 = \begin{pmatrix} 1 \\ 1 \\ 1 \end{pmatrix}$, $\mathbf{u}_2 = \begin{pmatrix} -1 \\ 0 \\ 1 \end{pmatrix}$, $\mathbf{u}_3 = \begin{pmatrix} -1 \\ 2 \\ -1 \end{pmatrix}$, that is multiples of vectors in (i)

and (ii).

17. (i) $\mathbf{u}_1 = \begin{pmatrix} 1 \\ 1 \\ 0 \\ 0 \end{pmatrix}$, $\mathbf{u}_2 = \begin{pmatrix} 1 \\ -1 \\ 2 \\ 0 \end{pmatrix}$, $\mathbf{u}_3 = \begin{pmatrix} -1 \\ 1 \\ 1 \\ 0 \end{pmatrix}$;

(ii) $\left\{ \begin{pmatrix} 1/\sqrt{2} \\ 1/\sqrt{2} \\ 0 \\ 0 \end{pmatrix}, \begin{pmatrix} 1/\sqrt{6} \\ -1/\sqrt{6} \\ 2/\sqrt{6} \\ 0 \end{pmatrix}, \begin{pmatrix} -1/\sqrt{3} \\ 1/\sqrt{3} \\ 1/\sqrt{3} \\ 0 \end{pmatrix} \right\}.$

Chapter 6

1. (i) (a) $\begin{pmatrix} 1/2 & -\sqrt{3}/2 \\ \sqrt{3}/2 & 1/2 \end{pmatrix}$, (b) $\begin{pmatrix} 1/2 & \sqrt{3}/2 \\ \sqrt{3}/2 & -1/2 \end{pmatrix}$, (c) $\begin{pmatrix} 1 & 0 \\ 3 & 1 \end{pmatrix}$;

(ii) (a) $(1/2, \sqrt{3}/2)$, $(-\sqrt{3}/2, 1/2)$; (b) $(1/2, \sqrt{3}/2)$, $(\sqrt{3}/2, -1/2)$;
(c) $(1, 3)$, $(0, 1)$.

2. (a) Rotation through $\pi/2$, matrix $\begin{pmatrix} 0 & -1 \\ 1 & 0 \end{pmatrix}$; (b) reflection in line through O at

120° with x-axis, matrix $\begin{pmatrix} -1/2 & \sqrt{3}/2 \\ \sqrt{3}/2 & 1/2 \end{pmatrix}$; (c) reflection in x-axis, matrix

$\begin{pmatrix} 1 & 0 \\ 0 & -1 \end{pmatrix}$; (d) the inverse of t, matrix $\begin{pmatrix} 1/2 & \sqrt{3}/2 \\ -\sqrt{3}/2 & 1/2 \end{pmatrix}$.

3. Eigenvalues $3, -2$, corresponding eigenvectors $\begin{pmatrix} 2 \\ 1 \end{pmatrix}$, $\begin{pmatrix} 1 \\ -2 \end{pmatrix}$.

5. $\begin{pmatrix} 0 & 1 \\ 1 & 0 \end{pmatrix}$; 45°; $y = -x$.

6. (a) $\lambda^2 - 5\lambda - 2 = 0$, (b) $\lambda^2 - 25 = 0$, (c) $\lambda^2 - 6\lambda + 16 = 0$, (d) $\lambda^2 - 9\lambda = 0$.

7. (i) $A\begin{pmatrix} 1 \\ -2 \\ 1 \end{pmatrix} = \begin{pmatrix} 3 \\ -6 \\ 3 \end{pmatrix}$, eigenvalue 3, $A\begin{pmatrix} 1 \\ 1 \\ 1 \end{pmatrix} = \begin{pmatrix} 6 \\ 6 \\ 6 \end{pmatrix}$, eigenvalue 6,

$A\begin{pmatrix} 1 \\ 0 \\ -1 \end{pmatrix} = \begin{pmatrix} -3 \\ 0 \\ 3 \end{pmatrix}$, eigenvalue -3.

(ii) Equation (6.7.2) becomes $\lambda^3 - 6\lambda^2 - 9\lambda + 54 = 0$, solutions $\lambda = 3, -3, 6$,

so respective solutions for equation (6.7.1) are $\begin{pmatrix} 1 \\ -2 \\ 1 \end{pmatrix}, \begin{pmatrix} 1 \\ 0 \\ -1 \end{pmatrix}, \begin{pmatrix} 1 \\ 1 \\ 1 \end{pmatrix}$.

8. $p \sim \begin{pmatrix} 2 & 0 \\ 0 & 2 \end{pmatrix}$; $q \sim \begin{pmatrix} 0 & -1 \\ 1 & 0 \end{pmatrix}$; $r \sim \begin{pmatrix} -1 & 0 \\ 0 & 1 \end{pmatrix}$; $s \sim \begin{pmatrix} 0 & 1 \\ 1 & 0 \end{pmatrix}$;

$t \sim \begin{pmatrix} 1/\sqrt{2} & -1/\sqrt{2} \\ 1/\sqrt{2} & 1/\sqrt{2} \end{pmatrix}$; $u \sim \begin{pmatrix} 1 & 3 \\ 0 & 1 \end{pmatrix}$.

9. $qr \sim \begin{pmatrix} 0 & -1 \\ 1 & 0 \end{pmatrix} \begin{pmatrix} -1 & 0 \\ 0 & 1 \end{pmatrix} = \begin{pmatrix} 0 & -1 \\ -1 & 0 \end{pmatrix}$, reflection in line $y = -x$;

$rs \sim \begin{pmatrix} -1 & 0 \\ 0 & 1 \end{pmatrix} \begin{pmatrix} 0 & 1 \\ 1 & 0 \end{pmatrix} = \begin{pmatrix} 0 & -1 \\ 1 & 0 \end{pmatrix}$, rotation about O through $90°$;

$sr \sim \begin{pmatrix} 0 & 1 \\ 1 & 0 \end{pmatrix} \begin{pmatrix} -1 & 0 \\ 0 & 1 \end{pmatrix} = \begin{pmatrix} 0 & 1 \\ -1 & 0 \end{pmatrix}$, rotation about O through $-90°$;

$st \sim 1/\sqrt{2} \begin{pmatrix} 0 & 1 \\ 1 & 0 \end{pmatrix} \begin{pmatrix} 1 & -1 \\ 1 & 1 \end{pmatrix} = \begin{pmatrix} 1/\sqrt{2} & 1/\sqrt{2} \\ 1/\sqrt{2} & -1/\sqrt{2} \end{pmatrix}$, reflection in line through

O at an angle $22.5°$ with x-axis;

$tr \sim 1/\sqrt{2} \begin{pmatrix} 1 & -1 \\ 1 & 1 \end{pmatrix} \begin{pmatrix} -1 & 0 \\ 0 & 1 \end{pmatrix} = \begin{pmatrix} -1/\sqrt{2} & -1/\sqrt{2} \\ -1/\sqrt{2} & 1/\sqrt{2} \end{pmatrix}$, reflection in line

through O at an angle $112.5°$ with x-axis.

11. (i) $\lambda^2 - \lambda - 2 = 0$, $\lambda = 2 \sim x = 2y$; $\lambda = -1 \sim y = -x$.
(ii) $\lambda^2 - \frac{6}{5}\lambda + 1 = 0$, no real eigenvalues, no fixed lines, rotation.
(iii) $(\lambda - 1)^2 = 0$, $\lambda = 1 \sim x = 0$, shear.
(iv) $\lambda^2 - 1 = 0$, $\lambda = 1 \sim x = 2y$; $\lambda = -1 \sim y = -2x$, reflection.
(v) $\lambda^2 - 7\lambda = 0$, $\lambda = 0 \sim x = -3y$; $\lambda = 7 \sim y = 2x$, collapse onto line $y = 2x$.

12. (i) $(\lambda - 1)(\lambda^2 - 2\lambda + 2) = 0$; only one real eigenvalue, $\lambda = 1 \sim x = y = 0$ (z-axis), rotation about z-axis.
(ii) $(\lambda - 1)(\lambda + 2)(\lambda - 4) = 0$, $\lambda = 1 \sim x = y = z$;
$\lambda = -2 \sim x = z = y/(-2)$; $\lambda = 4 \sim y = 0, x = -z$.

Chapter 7

1. $(10/3, 4/3, 1/3)$, $\mathbf{x}' - \mathbf{x} = \frac{4}{3}\mathbf{n}$, midpoint of XX' is $(8/3, 2/3, -1/3)$ whose coordinates satisfy the equation of the plane.

2. $(6, -1)$.

3. $\lambda = (\mathbf{x} - \mathbf{a}).\mathbf{u}$.

4. (i) $\mathbf{v}.\mathbf{v}' = r^2 \cos\theta$, (ii) $\mathbf{v} \times \mathbf{v}' = r^2 \sin\theta\mathbf{u}$.

6. (i) $7\mathbf{i} + 4\mathbf{j} - 5\mathbf{k}$, $-3\mathbf{i} + 24\mathbf{j} + 15\mathbf{k}$,
(ii) $0, 0, 0$,
(iii) $\mathbf{u}.(\mathbf{a} \times \mathbf{u}) = 0 = \mathbf{u}.(\mathbf{u} \times (\mathbf{a} \times \mathbf{u})) = (\mathbf{a} \times \mathbf{u}).(\mathbf{u} \times (\mathbf{a} \times \mathbf{u}))$.

7. (i) $-b\mathbf{i} + a\mathbf{j}$, (ii) same as (i).

8. $(2, 1/2 + \sqrt{3}/2, 1/2 - \sqrt{3}/2)$.

10. (i) (a) $(-p, q, r)$, (b) $(p, -q, r)$, (c) $(-p, -q, r)$, (ii) $(-p, -q, r)$, (iii) two successive reflections are equivalent to a rotation.

12. (i)
$$\begin{pmatrix} 1 - 2n_1^2 & -2n_1 n_2 & -2n_1 n_3 \\ -2n_1 n_2 & 1 - 2n_2^2 & -2n_2 n_3 \\ -2n_1 n_3 & -2n_2 n_3 & 1 - 2n_3^2 \end{pmatrix};$$

(ii) $-\begin{pmatrix} u_1^2(1 - \cos\theta) + \cos\theta & u_1 u_2(1 - \cos\theta) - u_3 \sin\theta & u_1 u_3(1 - \cos\theta) + u_2 \sin\theta \\ u_1 u_2(1 - \cos\theta) + u_3 \sin\theta & u_2^2(1 - \cos\theta) + \cos\theta & u_2 u_3(1 - \cos\theta) - u_1 \sin\theta \\ u_1 u_3(1 - \cos\theta) - u_2 \sin\theta & u_2 u_3(1 - \cos\theta) + u_1 \sin\theta & u_3^2(1 - \cos\theta) + \cos\theta \end{pmatrix}.$

Chapter 8

2. $\mathbf{r} = (a \cos\theta)\mathbf{i} + (b \sin\theta)\mathbf{j}$, $d\mathbf{r}/d\theta = (-a \sin t)\mathbf{i} + (b \cos t)\mathbf{j}$.

3. $\dot{\mathbf{r}} = (\cos t - t \sin t)\mathbf{i} + (\sin t + t \cos t)\mathbf{j} + \mathbf{k}$,
   $\ddot{\mathbf{r}} = (-2 \sin t - t \cos t)\mathbf{i} + (2 \cos t - t \sin t)\mathbf{j}$.
   The locus is a spiral lying on a cone with axis along the z-axis, and semi-vertical angle $\pi/4$.

5. $(1 + 2t - 3t^2)/\sqrt{3}$, $(2 - 6t)/\sqrt{3}$.

6. $\dfrac{b \sin\theta}{\sqrt{a^2 + b^2}}\mathbf{i} - \dfrac{b \cos\theta}{\sqrt{a^2 + b^2}}\mathbf{j} + \dfrac{a}{\sqrt{a^2 + b^2}}\mathbf{k}$.

7. If $b = 0$ the locus is of a circle radius a, so curvature $\kappa = 1/a$ and $\tau = 0$ since a circle lies in a plane.

8. $d\mathbf{r}/d\theta = 3 \cos\theta \sin\theta \, \dot\theta(-\cos\theta\mathbf{i} + \sin\theta\mathbf{j})$. This is the zero vector if $\theta = 0, \pm\pi/2, \pm\pi$, other values obtained by direct substitution.

9. $d\mathbf{r}/d\theta = -2 \cos\theta \sin\theta \, \dot\theta(\mathbf{i} - \mathbf{j})$, unit tangent vector at every point is $(\mathbf{i} - \mathbf{j})/\sqrt{2}$, so locus is a straight line $(x + y = 1)$.

10. (i) $2x\mathbf{i} + 2\mathbf{j}$, e^x.

11. (i) $2t\mathbf{i} + 2\mathbf{j}$, $(\cos t)\mathbf{i} + (\sin t)\mathbf{j}$,
    (ii) $t^2 \sin t - 2t \cos t - 5$,
    $(2t - 5 \cos t)\mathbf{i} + (-5 \sin t - t^2)\mathbf{j} + (-t^2 \cos t - 2t \sin t)\mathbf{k}$,
    (iii) $(t^2 - 2) \cos t + 4t \sin t$,
    $(5 \sin t + 2)\mathbf{i} - (5 \cos t + 2t)\mathbf{j} + \{(t^2 - 2) \sin t - 4t \cos t\}\mathbf{k}$.

12. (i) (a) $\dfrac{2}{t^2 + 2}\mathbf{i} + \dfrac{2t}{t^2 + 2}\mathbf{j} + \dfrac{t^2}{t^2 + 2}\mathbf{k}$, (b) $\dfrac{-2t}{t^2 + 2}\mathbf{i} + \dfrac{(2 - t^2)}{t^2 + 2}\mathbf{j} + \dfrac{2t}{t^2 + 2}\mathbf{k}$,

(c) $\dfrac{t^2}{t^2 + 2}\mathbf{i} - \dfrac{2t}{t^2 + 2}\mathbf{j} + \dfrac{2}{t^2 + 2}\mathbf{k}$, (d) $\dfrac{2}{(t^2 + 2)^2}$, $\dfrac{1}{(t^2 + 2)^2}$;

(ii) $1/18$, $1/36$.

Chapter 9

2. (i) Sphere, centre $(0, 0, 0)$, radius 2, (ii) the half-plane containing $(-1, 0, 0)$ and the z-axis, (iii) the plane $z = 0$.

3. (i) Line through $(-2, 0, 0)$ parallel to z-axis, (ii) circle, centre $(0, 0, 0)$, passing through $(2, 0, 0)$, (iii) half-line from $(0,0,0)$ through $(-1, 0, 0)$.

4. (i) Cylinder, axis along z-axis, radius 2, (ii) the half-plane containing $(0,1,0)$ and the z-axis, (iii) the plane $z = -3$.

5. (i) Line through $(0,2,0)$ parallel to the z-axis, (ii) circle, centre $(0,0,-3)$, radius 2, in plane $z = -3$, (iii) half-line from $(0,0,-3)$ through $(0,1,-3)$.

6. (i) Sphere on AB as diameter, (ii) sphere on OC as diameter, where $OC = \mathbf{a} + \mathbf{b}$.

7. $f_x = 2xy + 2y^3 - y, f_y = x^2 + 6xy^2 - x, f_x(2,-2) = -22, f_y(3,-1) = 24$.

8. $2x - 6y + z + 6 = 0$.

9. $2x + 13y + 2z + 12 = 0$.

10. (i) The semi-circle on line segment joining $(0,0,-2)$ and $(0,0,2)$ as diameter and passing through the point $(\sqrt{2}, \sqrt{2}, 0)$,   (ii) the horizontal circle with centre $(0,0,3/\sqrt{2})$ and radius $3/\sqrt{2}$,   (iii) the half-line starting at the origin, and passing through the point $(1,1,1)$.

11. (i) The vertical line through $(\sqrt{3}/2, 1/2, 0)$, (ii) the horizontal circle with centre $(0,0,-1)$ and radius 2, (iii) the half-line starting from $(0,0,2)$ and passing through $(1/2, \sqrt{3}/2, 2)$.

12. $f_x = 2xy - 2y, f_y = x^2 - 2x + 3y^2; 11y - z + 28 = 0$.

13. (i) $F_x = 2xyz + 2y^3$, $F_y = x^2z + 6xy^2 + 8yz^2$, $F_z = x^2y + 8y^2z$;

    (ii) $4x + y - 7z + 4 = 0$; (iii) $\dfrac{4}{\sqrt{66}}\mathbf{i} + \dfrac{1}{\sqrt{66}}\mathbf{j} - \dfrac{7}{\sqrt{66}}\mathbf{k}$.

14. $\nabla\phi = e^x\{(y^2 - yz)\mathbf{i} + (2y - z)\mathbf{j} - y\mathbf{k}\}, (10\mathbf{i} + 7\mathbf{j} - 2\mathbf{k})e$.

15. (i) $8xy^2$, (ii) $(\cos z - 3x^2)\mathbf{i} + (9x^2y - \sin z)\mathbf{j} + 2y(y^2 - x^2)\mathbf{k}$.

Index

Addition
  associative, 9
  commutative, 7, 9
  vector, 6–9
Associative laws, addition, 9
Axes, systems of, 3

Baked bean tin method, 38–40
Ball, definition, 112
Bases
  definition, 57, 63
  orthonormal, 62
  standard, 57
  theorems, 57–8
  for vector spaces, 57–9
  see also Orthogonal bases
Binormal vectors, definition, 101

Cartesian equations
  of planes, 31
  of straight lines, 11–13, 21
Centroids, and vectors, 18–19
Ceva, Giovanni (1648–1734), theorem, 22
Characteristic equation, definition, 79, 82
Column vectors, definition, 5
Commutative laws, addition, 7, 9
Components, definition, 4, 27, 35
Coordinate systems, non-rectangular, 106–21
Coordinates
  and vectors, 2
  see also Polar coordinates
Cosine rule, 29–30, 35
Curl
  concept of, 117–19
  definition, 119, 120
Curvature
  definition, 101, 104
  measurement of, 102
  radius of, 101, 104

Curves
  in three-dimensional space, 98–9, 103
    Serret–Frenet equations for, 100–3
Cylindrical polar coordinates, in three-dimensional space, 109–11, 119

Del, definition, 118
Derived vectors, definition, 96
Determinants, definition, 39
Differentiation
  partial, 112–14
  of vectors, 95–8
  rules, 99–100
Dimensions, definition, 58, 63
Direction cosines, definition, 6, 29
Directional derivative, definition, 118
Divergence
  concept of, 117–19
  definition, 118, 120
Dot product, use of term, 23

Eigenvalues, 72–8
  definition, 73, 82
  theorem, 73
Eigenvectors, 72–8
  definition, 73, 82
  theorem, 73
Enlargements
  and eigenvalues, 78
  in three-dimensional space, 80
  in two-dimensional planes, 69

Fixed lines, definition, 72
Free vectors, definition, 2
Functions, vector-valued, 94–105

Gradient
  concept of, 117–19
  definition, 118, 120
Gram–Schmidt orthogonalisation process, 60–2